Lecture Notes in Control and Information Sciences

Edited by A.V. Balakrishnan and M. Thoma

33

Peter Dransfield

Hydraulic Control Systems – Design and Analysis of Their Dynamics

Springer-Verlag
Berlin Heidelberg New York 1981

Author
Peter Dransfield, Ph.D.,
Department of Mechanical Engineering
Monash University, Australia

ISBN 3-540-10890-4 Springer-Verlag Berlin Heidelberg New York
ISBN 0-387-10890-4 Springer-Verlag New York Heidelberg Berlin

Printing and binding: Beltz Offsetdruck, Hemsbach/Bergstr.
2061/3020-543210

PREFACE

The Text centres around the ideas that for hydraulic control systems

- the quality of their dynamic response is important to their proper operation,
- prediction of dynamic response capability is best obtained via the development of dynamic models,
- the inability of many designers to efficiently effect predictive dynamic analysis has its roots in a lack of confidence in the forming of models,
- the use of *power bond graphs* is a natural and superior approach to the development of reliable dynamic models,
- digital simulation provides the most appropriate method for extracting response information from dynamic models.

Among these, the use of bond graphs (abbreviation for power bond graphs) is paramount to the Text. Therefore, bond graphs are introduced, explained, demonstrated, and utilized in substantial detail. The features of bond graphs which make them desirable in the hydraulic control system design situation include their structural affinity with the real system and its components, the formality in the assembly of a bond graph structure, their modular and re-usable nature, and the provision of a structure from which the set of equations which become the model can be formally prepared.

It is assumed that the reader has a general appreciation of the nature and operation of hydraulic control systems and their major components. The Text is written mainly for those mechanical and control engineers who would like to be able to include predictive dynamic analysis among their techniques for the design of hydraulic control systems but who find existing approaches unsatisfying or inappropriate. The Text should be suitable also for up to twenty hours in appropriate graduate or senior undergraduate courses, and for specialized extension-type courses.

For full appreciation of the Text, the reader will need to make a "break through" by thoroughly understanding and using the symbols and structures of bond graphs. Such a break through was literally forced on me by a graduate student whom I was supervising for a research Master degree. I gladly acknowledge my debt to Bevis Barnard, Senior Lecturer at Caulfield Institute of Technology, for the gentle persuasion emanating from his own discovery and conviction that bond graphs offered something unique to the modelling, simulation and design of powered control systems. I acknowledge also the conceptions of Henry Paynter, the original and sustained developmental work of Dean Karnopp and Ron Rosenburg, the work of bond graph entrepreneur Jean Thoma, the assistance of colleague Jacek Stecki, and the contributing work of a small but steady stream of undergraduate and graduate students including Rob Winton, M.K. Teo, Roger LaBrooy, Kevin Duke, S. Ramachandran, V.K.L. Mai, and Merren Cliff.

Many people contribute to the production of a text. I single out two for special thanks and appreciation; Lorry Ryan and John Millar, both of Monash University, for their excellent work in typing and drafting respectively.

And there is an individual without whose encouragement this Text would not have been completed —— my wife Nevillie whose support was sustained, inspirational, and essential.

To those willing to learn, and to those willing to try a new way, welcome aboard.

P.D. March 1981

CONTENTS

CHAPTER 7 (contd.)

CHAPTER 1

INTRODUCTION

1.1 INTRODUCTION

Hydraulic control systems are used to control the position or speed of resisting loads. Final drive is usually provided by a hydraulic actuator —— either a linear-motion hydraulic cylinder or a rotary-motion hydraulic motor. The actuator develops its force or torque by receiving liquid from a positive displacement pump at a relatively high pressure, usually between 7 and 35 MPa. The actuator develops its motion by receiving flowrate of liquid from the pumping system. Fig. 1.1 illustrates several simple hydraulic control systems.

Hydraulic control systems are used

- where relatively large forces or torques are required (industrial presses; mobile lifting, digging, and materials handling equipment, etc.),
- where fast, stiff response of resisting loads is required (machine tool drives, flight simulators, etc.),
- where close control of response is required (aircraft control surfaces, machine tool slides, industrial robots, etc.),
- where manual control over substantial powered motions is required (mobile equipment, aircraft controls, road vehicle steering, etc.),
- as the final actuator sub-system in complex automatically controlled situations (electrohydraulic flight simulators, industrial robots, fatigue and other programmable test rigs, etc.).

a. A Linear Drive Hydraulic Control System

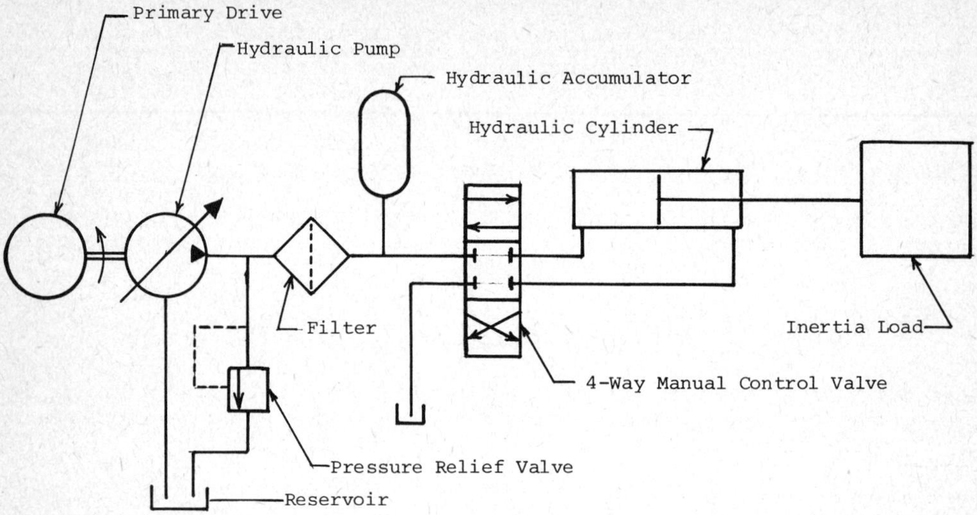

b. A Rotary Drive Hydraulic Control System

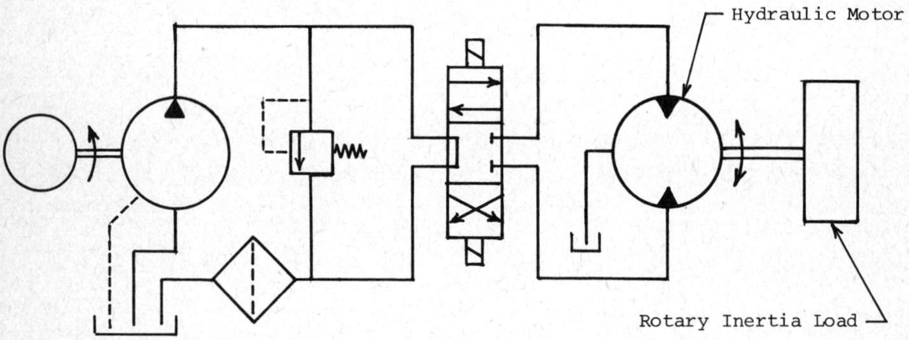

FIG. 1.1 TYPICAL HYDRAULIC CONTROL SYSTEMS

That is, they are used in dynamic situations —— situations where response to commands can't be instantaneous, but must be controlled and very often optimized to follow pre-determined motion-time paths. Fig. 1.2 illustrates two simple command and response pairs.

It is not sufficient for the designer of a hydraulic system to know that his proposed system *will* move the driven load from one state to another. He should know also *how* the load will move. He needs to appreciate not only the initial and final states of the responses, but also the time-domain path between these states (Fig. 1.2). He should know if system response is stable, if it is fast enough or too fast, if it is oscillatory or aperiodic, etc.

Dynamic analysis can be

- corrective in nature, the objective being to find out why an existing system is not performing satisfactorily and to deduce what can be done to improve the situation,

- synthetical in nature, with the objective of ensuring good dynamic response before commitment to system hardware.

In the latter situation, the proposed system's dynamics are designed as part of the system design process. The present Text is mostly concerned with this approach, though it is relevant also to corrective analyses.

1.2 A DESIGN PHILOSOPHY

Fig. 1.3 illustrates the stages in a dynamic design philosophy. *Specify Task* does not imply use of any particular system or technology —— simply a definition of the task to be accomplished. *Define Required Response* is also a consideration independent of the particular system or technology to be employed —— typically this would be a time-based curve of the desired system response form, with time and response magnitudes as explicit scales, such as is shown on Fig. 1.2.

The stage *Propose System* requires the first commitment to a technology —— in the present Text a decision to use a hydraulic control system, whose form is then proposed. Initial selection of the hardware is decided usually by a combination of

- past experience,
- the maximum pressure to be allowed,
- static power and overall motion considerations,
- the available hardware components, and their cost.

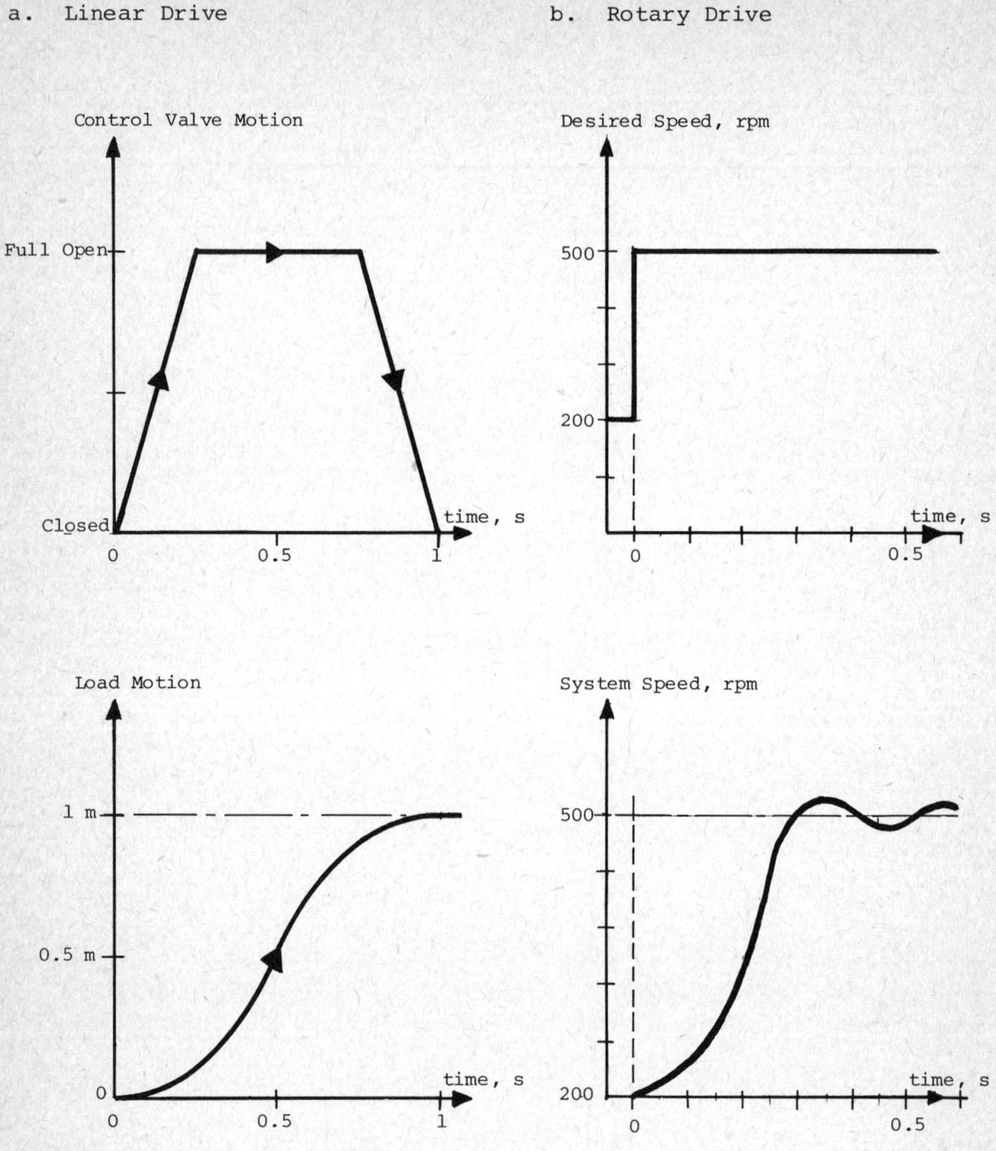

FIG. 1.2 SOME COMMAND-AND-RESPONSE PAIRS

(Response times can be much faster or slower than those shown, depending on system and load and input)

It is after this initial decision that a more conclusive and enlightening dynamic response analysis should be made, requiring the steps *Develop Dynamic Model* and *Predict Response* (by solving the model). The predicted response is compared with the desired response (*Compare*) and the discrepancy between them is assessed (*Judge Response Quality*).

A decision is made then that the performance of the proposed system is

- adequate (*Response Acceptable; Finalize Design*),
- encouraging enough so that simple parameter adjustments will make it acceptable (*Response Promising; Optimize It*),
- relatively hopeless, requiring re-design of the proposed system (*Response Hopeless; Re-design System*).

Dynamic performance analyses are often carried out for the high-cost hydraulic systems associated with aircraft, military, space, and some machine tool applications. They are less often associated with the design of industrial systems. One result is that while many such systems are accepted by the user, their performance rarely approaches a considered optimum.

Relatively few mechanical (or other!) engineers are confident and competent with predictive dynamic analysis of systems. Predictive dynamic analysis hinges on the formation of a mathematical representation of the system. This mathematical model is used to extract information on the response characteristics of the system. If an adequate mathematical model can't be formed quickly and with confidence, dynamic analysis is not an attractive proposition to most designer-analysts.

Few designers of hydraulic control systems are able to confidently form dynamic models of proposed systems. On the other hand, given a mathematical model, most engineers can arrange for its solution, which provides the required prediction of dynamic performance. Generally, solution of anything other than the simplest models is best effected by digital computer. Digital simulation procedures, widely available in easy-to-use forms, are of particular relevance. There is no need for the designer-analyst to be himself a computer programmer. Given the means to solve (simulate) his model, the designer-analyst can change one or more of the model's parameters in order to seek a desired form of system response. Such optimization of response can be sought through repetitive solutions of the model as parameters are changed. Automated optimization procedures can also be used.

Accepting that solution of mathematical models is relatively easy, the great hurdle for the designer-analyst lies in the formation, with speed and confidence, of

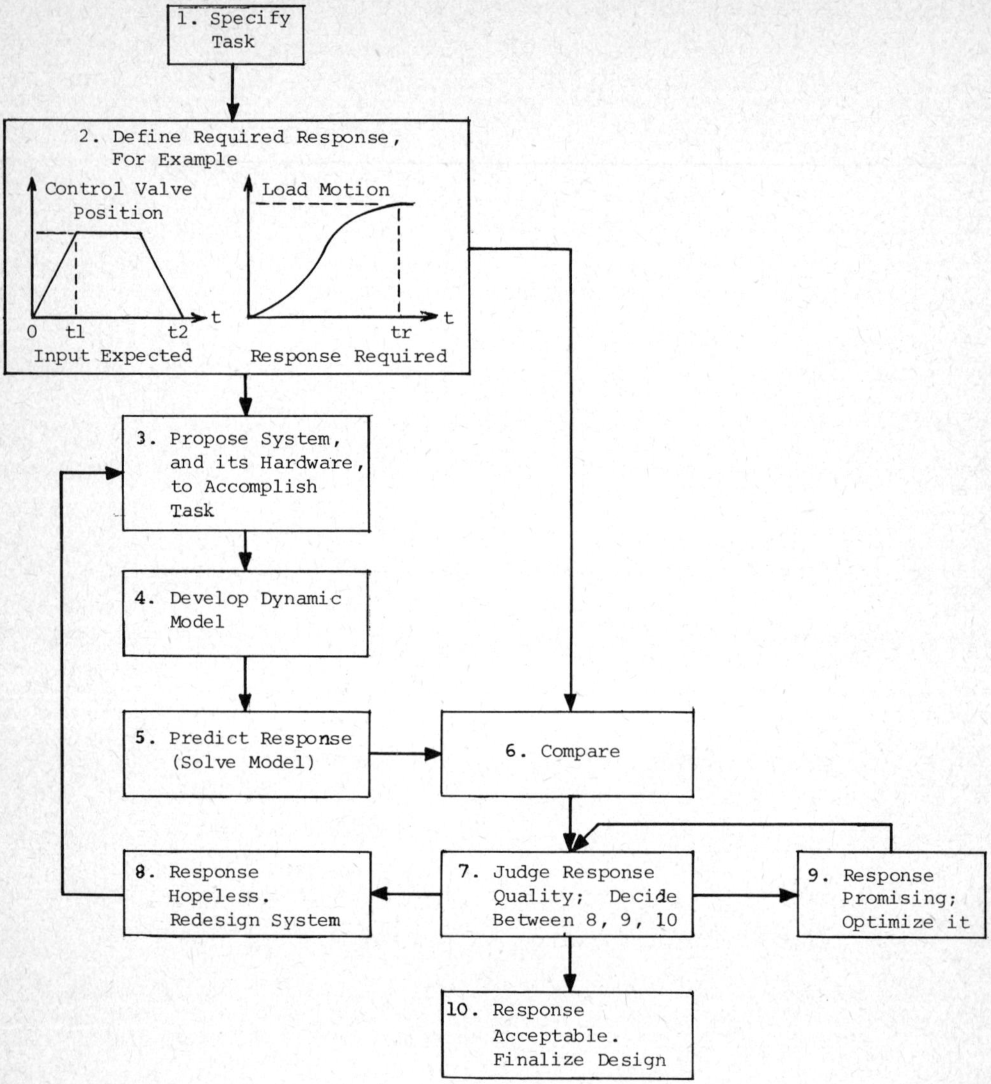

FIG. 1.3 APPROACH FOR DESIGNING DYNAMIC RESPONSE

an adequate mathematical model of his proposed system.

1.3 DYNAMIC MODELLING

Conventional dynamic modelling procedures for control systems, such as

- transfer functions,
- block diagrams and signal flow graphs,
- sets of equations not having a unique formal arrangement,
- vector-matrix (state space) expressions,

have been around for a long time. Journals contain many examples of successful applications of each of them. Yet the lack of general use of these model forms by designers of hydraulic systems suggests that they are insufficiently attractive.

It appears that the conventional dynamic modelling procedures do not adequately provide for some of the designer's needs, including

- the need for easy-to-follow formality in the modelling process,

- the need for reliable re-useable models of common hydraulic system components. Hydraulic control systems are of modular construction, being made up by assembling and interconnecting a set of well-defined off-the-shelf hardware devices. A particular type and size of pump, valve, actuator, etc. is used in many different systems. The designer needs to have available a set of dynamic models, one for each common hydraulic system component, which is capable of being interconnected to form a system model in a manner reflecting the formation of the hardware system itself,

- the need to be able to change easily the model of a proposed system to reflect conceivable variations in the proposed system, aimed at ensuring choice of the best design for the particular system.

An engineering designer needs to be creative —— to have imagination —— for the conception of design solutions. For subsequent analyses however, he requires availability of well-established and readily-applied techniques.

Hydraulic control systems have some distinctive features relating to dynamic analysis, including the following:

They are usually of relatively small scale. In terms of a transfer function, their dynamics often can be described adequately by a second, third, or fourth order differential equation. Rarely do they require more than sixth order transfer functions, although complex multi-axis systems occur. Also, they are often compact and amenable to a lumped parameter approach. Thus, they are potentially amenable to dynamic modelling;

- They are inherently non-linear due to the presence of turbulent orifice flow, friction, valve lap, pressure limiting actions, and other effects;

- They usually are expected to provide large scale response. Quite normally they are commanded to move the controlled load from one extreme position to the other;

- These latter two factors cause linear analysis techniques to be of very limited use;

- They are power transducing, power transmitting systems. Typically, power is drawn from a source such as an electric mains supply, diesel engine, aircraft engine, etc., and progressively converted and transmitted to a mechanical form useable at the driven load.

Conventional modelling procedures do not particularly recognize these factors, nor particularly utilize them to advantage. The reason-for-being of this Text is the proposition that dynamic power exchanges taking place between system components provide the key for incorporating dynamic response analysis into the system design environment. *POWER FLOW MODELLING* procedures allow a dynamic description of the system which retains a close association with the physical connections and actions of the modular system itself.

Power flow modelling, and in particular its recent extension *POWER BOND GRAPHS*, provides the designer of powered control systems with a dynamic modelling procedure which is more fundamental to, and more useable in, the system design environment. It is intended to demonstrate this by the end of the Text. It is intended also to leave the reader with a library of power bond graph models of the more common hydraulic system components, and with the ability to develop power bond graph models of any additional components he might wish to use.

Fig. 1.4(a) illustrates a simple hydraulic system, Fig. 1.4(b) a power bond graph structure for the system, and Fig. 1.4(c) the set of equations which is a dynamic model of the system. The bond graph structure is the key for linking the system to its dynamic model. The bond graph is a type of signal flow diagram which to a person familiar with bond graphs bears a striking affiliation with the hardware system diagram. The set of equations is readily formed from the bond graph structure.

Each line in the horizontal chain of Fig. 1.4(b) represents a physical component through which power flows between the adjacent system components. The product of

a. The System

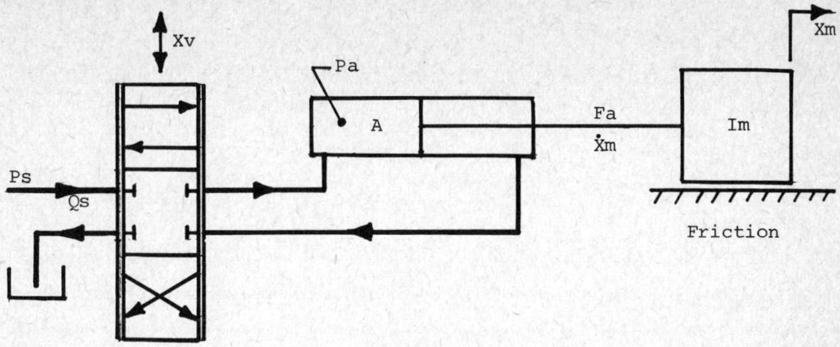

b. Power Bond Graph
(Return flow neglected, Xv positive only)

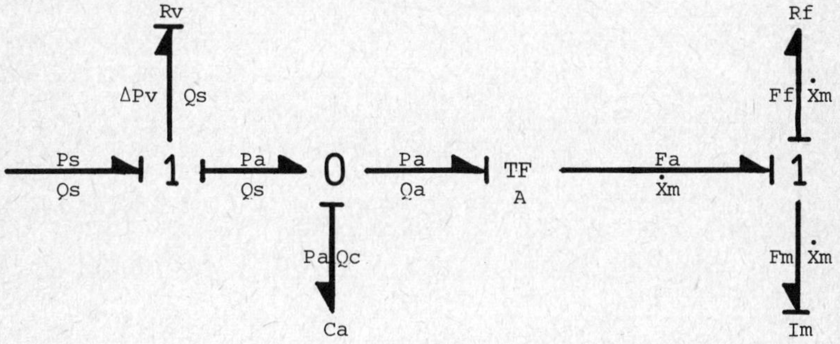

c. Dynamic Equations

$$Ps = \text{constant (assumption)}$$

$$Qs = K \cdot Xi \cdot \Delta Pv^{\frac{1}{2}} \quad \text{(for Xv positive)}$$

$$Pa = Pa(o) + 1/Ca \cdot \int Qc \, dt \quad \text{(linear capacitance equation)}$$

$$Ff = R(Xa) = \text{friction force for cylinder}$$

$$Fm = Fa - Ff$$

$$\dot{X}m = \dot{X}(o) + 1/Im \cdot \int Fm \, dt \quad \text{(Newton's law)}$$

$$Fa = Pa \cdot A$$

$$Qa = \dot{X}m/A$$

$$\Delta Pv = Pv - Pa$$

$$Qc = Qs - Qa$$

FIG. 1.4 AN ELEMENTARY HYDRAULIC CONTROL SYSTEM AND ITS BOND GRAPH MODEL

the two variables written against each line describes the power flowing in the line. Thus Ps • Qs (supply pressure times supply flowrate) describes the power flowing into the control valve. Similarly, Fa • $\dot{X}$m (force applied to the mass times velocity of the mass) describes the power flowing into the driven load. The R , C , and I symbols appearing at the ends of the spur lines of the diagram indicate respectively the presence of resistive (R) , capacitive (C) , and inertive (I) effects which affect the system's dynamic response. The half-arrows on each line indicate the power flow direction in each power bond, either real or assumed. The short transverse bar at one end of each line indicates a cause-effect decision aimed at arranging equations, to be formed later from the bond graph structure, in a form most suitable for simulation. These various points will be developed and explained in the Text. The Text's nomenclature and units are discussed in Chapter 2.

Formation of the model structure, i.e. of the power bond graph, decisions on the precise relationships to be used in the equations, and formation of and arrangement of the equation set, are separate and progressive steps in the dynamic model derivation procedure. In conventional modelling procedures these steps become jumbled together, often creating confusion in the minds of part-time dynamic analysts. It will be demonstrated in later chapters that power flow models allow the formality and flexibility required in the system design situation, particularly in that

- component models are useable in a variety of system models, as are the hardware components themselves useable in a wide variety of systems,
- system models are readily formed by direct coupling of component models, as are systems formed from the components,
- local changes in model structures are easy to make,
- model relationships can be varied locally without disrupting the remainder of the model.

Power flow and power bond graph modelling procedures can be used with all types of powered control systems —— electrical, pneumatic, mechanical, etc. The present Text however concentrates on their use with hydraulic systems, although this encompasses electric motors, electrohydraulic devices, and a variety of mechanical sub-systems. The ideas of power flow and power bond graph modelling have been developed by others, and references will be provided in appropriate places. The present Text simply utilizes power flow techniques and orientates them specifically towards use in the design and analysis of hydraulic control systems.

Beyond a review in Appendix 1, with References, the present Text does not provide instruction on how to develop a basic design for a hydraulic system for a specific application. It concentrates on the inclusion of dynamic response analysis as part of the system design procedure.

1.4 DESIRABLE BACKGROUND

The user-reader of the Text ideally would have as background

- familiarity with basic hydraulic control system components, systems, and the symbols used to represent hydraulic components in system diagrams of the type of Fig. 1.1,
- familiarity with hydraulic system design to the extent of being able to select for specified applications the major components such as pumps, relief valves, control valves, and actuators,
- familiarity with engineering dynamics to at least the stage of appreciating common physical concepts and relationships such as Newton's law, the potential drop-flowrate-resistance concept, the potential-flowrate-capacitance or compliance concept, etc. It is basic to all dynamic modelling procedures that such actions and relationships be appreciated,
- familiarity with some of the conventional modelling procedures used in first-course control theory or dynamic systems analysis.

The Text is primarily aimed

- at practising designers of hydraulic control systems, and engineers who from time to time must design or specify the performance required of hydraulic systems,
- at persons interested in control engineering, automation, or design whose duties or vision encompasses possible use of hydraulic control systems,
- at students in specialized undergraduate courses, graduate courses, and extension courses, involving dynamic analysis and fluid power systems.

1.5 ARRANGEMENT OF TEXT

The Text is arranged as follows:

- Chapter 1 provides the setting for the Text and makes its case;
- Chapter 2 provides a rational Nomenclature which aids in easy recognition of symbols in equations. It also discusses the use of SI units and provides

a table of those used in the Text;

- Chapter 3 provides a review with examples of conventional modelling procedures used in dynamic analysis;
- Chapter 4 introduces and demonstrates the power flow modelling approach, providing the basis for developing power bond graph techniques;
- Chapter 5 introduces power bond graphs, the key feature of the Text, which is offered as a major step towards making predictive dynamic analysis viable for more designers of hydraulic control systems;
- Chapter 6 reviews briefly digital simulation of power flow models, taking the view that this is the most practical and most available approach to the solving of models;
- Chapter 7 discusses and demonstrates the choice of equation and relationship forms for physical actions and devices. It also discusses choice of numerical values for coefficients in the equations;
- Chapter 8 describes some applications of bond graph modelling and digital simulation of hydraulic systems;
- Chapter 9 briefly outlines response optimization techniques relevant to hydraulic system design;
- Chapter 10 briefly introduces some factors which can affect the performance of hydraulic systems, and about which the system designer should be aware;
- Chapter 11 is a summary and conclusion;
- Appendix 1 reviews briefly conventional (static) design approaches for specifying the components of hydraulic systems and their size ratings;
- Appendix 2 provides conversion factors for British to SI units, for the quantities used in the Text;
- Appendix 3 provides a limited bibliography on dynamic analysis of hydraulic control systems.

1.6 CONCLUSION

In conclusion, the orientation attempted with the Text can be stated: "Given a proposed hydraulic control system, simulate it to check and to optimize its dynamic performance *before* any final commitment is made to expensive hardware".

CHAPTER 2

NOTATION AND UNITS

2.1 INTRODUCTION

It is difficult enough to master reported dynamic analyses without the problem of inconsistent and non-standard notation and units. It is inevitable that many subscripted symbols will be used, which adds to the confusion. Various authors use British, metric, and Systèmes Internationales (SI) units, and sometimes mix dimensions within their chosen unit schemes.

What a help if everyone would use the same notation and units! This would allow immediate recognition of symbols, and of the significance of the magnitudes of coefficients and numbers appearing in relationships. The value of this kind of recognition in the system design environment should not be under-estimated. The designer needs to be able to quickly utilize established techniques and data, rather than seek out new unproven and possibly ambiguous material.

The notation, the way of writing relationships, and the units used in this Text seek standardization.

2.2 NOTATION

2.2.1 Symbols

Fig. 2.1 lists the symbols used. General criteria are:

- Upper case English letters are used generally for system variables and coefficients, each being reserved for particular quantities; some widely-

a. Parameters (Coefficients and Variables)

Capitals	Lower Case	Greek
A - area	a -	α - angular acceleration
B - bulk modulus of oil	b -	β -
C - capacitance, or capacitive	c -	γ -
effect, or compliance	d -	δ -
D - operator d/dt	e -	ε -
E - general symbol for effort	f -	ζ - damping ratio
(i.e. potential) variable;	g - gravitational	η - efficiency
also voltage	acceleration	θ - angular displacement
F - force	h -	ι
G -	i - electrical	κ
GY - gyration power transformation	current	λ
H -	j -	μ - coefficient of
I - inertive or inductive effect;	k -	friction
inertia or inductance	ℓ -	ν - kinematic viscosity
J -	m -	ξ -
K - general symbol for coefficient	n -	o -
L - length	o -	π -
LR - lever ratio	p -	ρ - density
M -	q -	σ -
N - revolutions per unit time	r -	τ - time constant
O -	s -	υ -
P - pressure	t - time	ϕ -
Q - volumetric flowrate; general	u -	χ -
flowrate variable	v -	ψ -
R - resistance, or resistive	w -	ω - angular velocity
effect	x -	
S - source; constant or	y -	
independent variable	z -	
SE - source of effort (potential)		
SQ - source of flowrate		
T - torque		
TF - direct power transformation		
U -		
V - volume		
W -		
X - linear displacement		
Y -		
Z -		

b. Suffices (all lower case)

a - actuator (cylinder or motor)	n -
b - filter, strainer	o - oil
c - fluid compliance (compressibility)	p - pump
d -	q -
e - exhaust (tank, or return line)	r - relief valve
f - friction	s - supply
g - gravity	t -
h - hose, or fluid line	u - coulomb friction
i - input	v - valve
j - rotating inertia	w -
k - spring, mechanical compliance	x -
ℓ - leakage	y -
m - linear motion inertia (mass)	z -

1, 2 ... 9 denote different local values

FIG. 2.1 STANDARD NOTATION

used Greek letters are retained; for gravitational acceleration, for time, and for electrical current, g , t and i respectively are retained;

. Subscripts are lower case English letters, each reserved for a particular significance; numbers are used also;

. Currently accepted usage and physical association of symbols with particular quantities are retained as far as is feasible;

. Subscripts are written *in line* with their parent symbol; multiple subscripts are allowed;

. It is assumed that the reader will know or will quickly realize which parameters are dynamic (time dependent) and which are constants. The symbology (t) to make the distinction is rarely used, e.g. P and not P(t) is used for a dynamic pressure. If a normally dynamic parameter is assumed to be constant, it is prefixed with S to denote Source, e.g. SP denotes a constant pressure.

2.2.2 Examples of Symbol Use

. Ck denotes compliance (C) of a spring (k) ;

. Bo denotes the bulk modulus (B) of oil (o) ;

. Ip denotes inertance (I) of pump (p) rotating parts;

. Rℓp denotes a resistance (R) expression, or a resistance coefficient, describing leakage (ℓ) from a pump (p) ;

. Kvl denotes a particular (1) valve (v) coefficient K , where there is more than one such coefficient present (Kv2 , etc.);

. θa denotes the angular displacement of a hydraulic motor (actuator).

2.2.3 Operations

. Multiplication is indicated by a central dot: N • P ; Np • Ck ; 13 • Xm ;

. Decimal point is a low dot: 9.8 ; 9.8 • Np ;

. Division is indicated by slash: N/P ; Np/Ck ; 9.8/Np ;

. Brackets are used only where necessary to prevent ambiguity: Ac/(Vc • Bo) ;

. Exponents appear in a normal raised position, placed immediately after the last subscript and raised just above it: X^2 ; Xm^2 ; $Xm3^2$; $Xm^{0.5}$;

- Integration is indicated by the usual $\int \ldots dt$, with one space, but no dot, before dt : $\int Pa\ dt$; $\int Pa \cdot C2\ dt$;

- The overdot system is used for time derivatives; linear velocity is $\dot{X}$; linear acceleration is $\ddot{X}$, etc.;

- Raising ten to a power is expressed 10 Ex or 10 E-x (10^x and 10^{-x} respectively).

2.2.4 Equations

The following are typical expressions appearing in this Text:

$$Q\ell p = Pp/R\ell p$$

$$Qp = Kv \cdot Xi \cdot \Delta P^{\frac{1}{2}}$$

$$\dot{X}m = \dot{X}m(0) + 1/Im \cdot \int Fm\ dt$$

$$Bo = 1750\ E6 \quad (\text{i.e. } 1{,}750 \times 10^6)\ \text{Pascal}$$

2.2.5 Comments

- A little usage will show that this scheme is easy to write, easy to type, easy to understand, and is easily associated with digital computer programming and print out.

- Some of the terms included in Fig. 2.1 apply specifically to power bond graph terminology to be introduced later.

- Only symbols used in this Text are designated in Fig. 2.1. There are plenty of gaps for future assignment or general use.

2.3 UNITS

Systèmes Internationales (SI) units are used without exception. Apart from its substantial and growing international acceptance, the SI system overcomes most of the ambiguity associated with the use in engineering and science of the British and the standard metric systems of units.

The SI units which appear in this Text are shown in Fig. 2.2. Appendix 2 provides some useful conversion data.

Physical Quantity	Name of Unit	Symbol of Unit	Conversational Alternatives
Basic Units			
Length	metre	m	centimetre (cm), millimetre (mm)
Mass	kilogram	kg	tonne (E3 kg)
Time	second	s	minutes
Electric current	ampere	A	
Temperature	degree celsius	C	
Derived Units			
Area	$metre^2$	m^2	
Volume	$metre^3$	m^3	litre (E-3 m^3)
Density	$kilogram/metre^3$	kg/m^3	
Force (also weight)	newton	N	kilonewton (kN)
Torque	newton metre	N · m	
Rotating inertia	kilogram $metre^2$	kg · m^2	
Pressure	pascal (N/m^2)	Pa	kPa, MPa, bar(E5 Pa)
Energy (work)	joule	J	
Power	watt	W	kilowatt (kW)
Electric potential (electric voltage)	volts	E	
Electric resistance	ohm	Ω	
Velocity	metre/second	m/s	
Acceleration	$metre/second^2$	m/s^2	
Flowrate	$metre^3/second$	m^3/s	litre/sec or litre/minute
Angle	radian	rad	revolution
Angular velocity	radian/second	rad/s	revolution/sec or minute
Frequency	hertz	Hz	cycle per second
Dynamic viscosity	newton $second/metre^2$ or pascal second	N · s/m^2, or Pa · s	centipoise (cP)
Kinematic viscosity	$metre^2/second$	m^2/s	centistoke (cSt)

Common multiples: micro (u) E-6 ; milli (m) E-3 ; centi (c) E-2 ; kilo (k) E3 ; mega (M) E6

FIG. 2.2 RELEVANT SI UNITS AND SYMBOLS

(See Appendix 2 for Conversion Factors)

2.3.1 Comments

- In conversation or in descriptive passages, convenient multiples of basic SI units can be useful. For example, flowrate in litres/second (ℓ/s) ; pressure in megapascals (MPa) ; force in kilonewtons (kN) ; angular velocity in revolution per minute (rpm) . The author considers the pressure unit *bar* to be an unnecessary bastardization of the SI system and discourages its use (1 bar = 10^5 Pa) .

- For numerical work (specifying coefficients, applying relationships, etc.), use *only* base SI units, for example, m^3/s for flowrate, Pa for pressure, N for force, and radian/second (rad/s) for angular velocity. Failure to do this can lead to gross and unsuspected magnitude errors in calculations, with 10 E ± n , n = 1 ... ∞ , floating around. One can easily enough talk or write descriptively of 10 MPa , but should write it as 10 E6 Pa when it comes to equations.

- For unit products use central dot to separate component units

  i.e. N · s , not Ns , for Newton second,

  kg · m^2 , not kgm^2 , for kilogram $metre^2$.

- Take care with unit exponents where slash is used

  i.e. $m^2 \cdot Pa^{-1} \cdot s^{-1} = m^2/(Pa \cdot s) \neq m^2/Pa \cdot s$

2.4 REFERENCES

For a more complete description of the S.I. system and its units see:

2.1 Blackman, D.R., SI units in Engineering, MacMillan, 1969.

2.2 Chiswell, B. and Grigg, E.C.M., SI Units, J. Wiley, 1971.

2.3 The International System of Units, Nat. Bur. Stand. (U.S.), special publication No. 330, 1972.

2.4 The ABC's of SI, National Fluid Power Association, USA, 1974.

CHAPTER 3

CONVENTIONAL MODELLING PROCEDURES

3.1 INTRODUCTION

- Transfer functions,
- block diagrams and signal flow graphs,
- vector-matrix (state-space) equations,
- informally arranged sets of dynamic and algebraic equation,

are regarded in this Text as conventional forms for dynamic models. They are conventional in that each is widely used in hydraulic system analyses readily available in texts and articles.

Readers familiar with the application of conventional modelling procedures to hydraulic control systems can pass over this chapter. Others may prefer to delve into it to whatever degree is needed for revision or orientation. The Chapter assumes familiarity with the above model forms, and simply demonstrates briefly their application as a prelude to development of power flow and power bond graph modelling in following chapters. The distinctions and advantages of the power flow procedures will be developed in the later chapters.

3.2 CONVENTIONAL MODEL FORMS

Fig. 3.1 a illustrates an inertia load Im with frictional damping Rf and support compliance Ck , driven by a hydraulic ram (Area A) which is controlled by a 4-way closed-centre valve (volumetric flow coefficient Kv1 , pressure flow coefficient Kv2). Load position feedback is provided through the mechanical linkage L1 , L2 , the assembly thus becoming a servomechanism with lever position

a. System Schematic

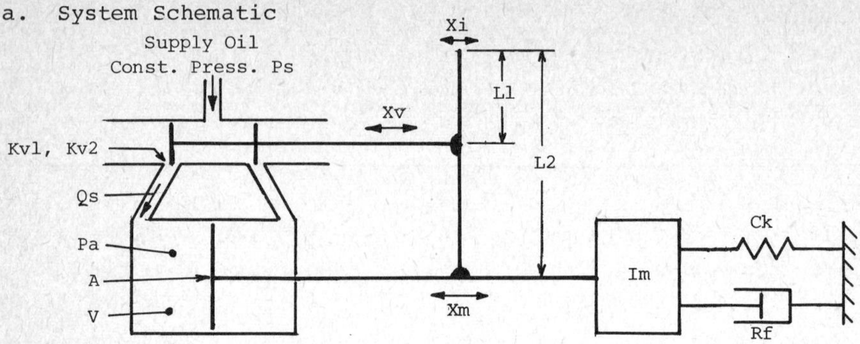

b. Equation Set Model

$$Xv = (L2 - L1)/L2 \cdot Xi - L1/L2 \cdot Xm \qquad 3.1$$

$$Qs = Kv1 \cdot Xv - Kv2 \cdot Pa \qquad 3.2$$

$$Qs = A \cdot \dot{X}m + V/B \cdot \dot{P}a \qquad 3.3$$

$$Im \cdot \ddot{X}m = Pa \cdot A - Rf \cdot \dot{X}m - Ck \cdot Xm \qquad 3.4$$

c. Transfer Function

$$\frac{Xm}{Xi} = \frac{K1}{K2 \cdot D^3 + K3 \cdot D^2 + K4 \cdot D + 1}$$

where

$$K1 = (L2 - L1)/[L1^2/L2 + L1 \cdot Kv2 \cdot Ck/(Kv1 \cdot A)]$$

$$K2 = V \cdot Im/[B \cdot Kv1 \cdot A \cdot L1/L2 + B \cdot Kv2 \cdot Ck]$$

$$K3 = [V \cdot Rf/B + Kv2 \cdot Im]/[L1/L2 \cdot Kv1 \cdot A + Kv2 \cdot Ck]$$

$$K4 = [A^2 + V \cdot Ck/B + Kv2 \cdot Rf]/[Kv1 \cdot A \cdot L1/L2 + Kv2 \cdot Ck]$$

d. Vector-Matrix Model, of form $\dot{\bar{X}} = \underline{A} \cdot \bar{X} + B \cdot \bar{u}$

$$\begin{vmatrix} \dot{X}m \\ \ddot{X}m \\ \dot{P}a \end{vmatrix} = \begin{vmatrix} 0 & 1 & 0 \\ -Ck/Im & -Rf/Im & A/Im \\ -L1/L2 \cdot Kv1 \cdot B/V & -A \cdot B/V & -Kv2 \cdot B/V \end{vmatrix} \begin{vmatrix} Xm \\ \dot{X}m \\ Pa \end{vmatrix} + \begin{vmatrix} 0 \\ 0 \\ (L2-L1)/L2 \cdot Kv1 \cdot B/V \end{vmatrix} \begin{vmatrix} Xi \end{vmatrix}$$

e. Block Diagram

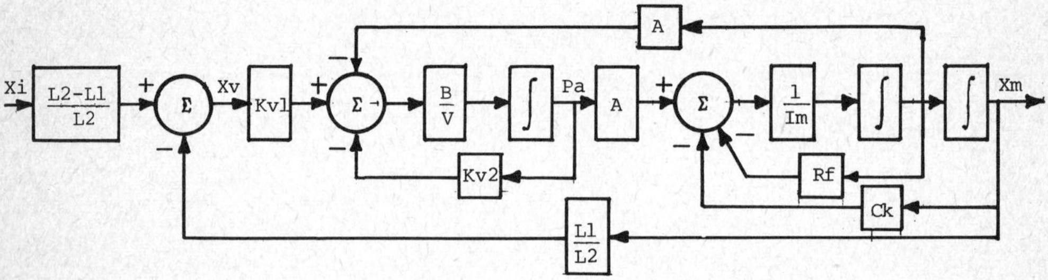

f. Signal Flow Graph

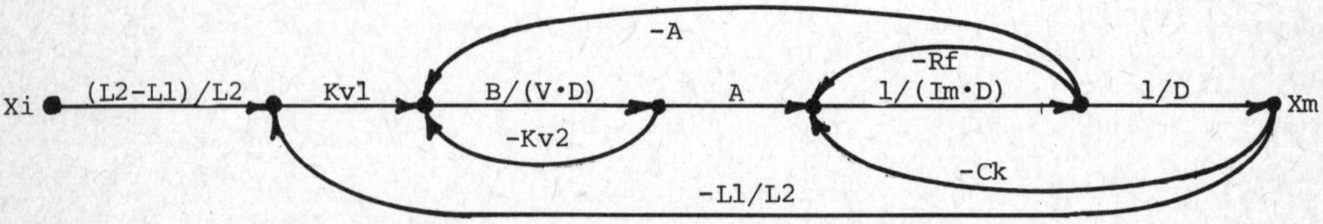

FIG. 3.1 SOME CONVENTIONAL MODEL FORMS FOR SYSTEM OF FIG. 3.1a

Xi as the controlling input and load position Xm as the controlled output. Supply pressure Ps is assumed to remain constant. The lever system L1 , L2 is such as to allow small control valve motions Xv and much larger load motions Xm . Xi is a small motion. This system will be used to demonstrate conventional model forms.

3.2.1 Informal Equation Set

A dynamic model may be in the form of a set of simultaneous equations, linear and non-linear, algebraic and differential, describing the relationships between the system variables and coefficients and the inputs. The arrangement of the equations is chosen by the analyst. For example, the dynamic response of the system of Fig. 3.1 a can be described by the equation set of Fig. 3.1 b , in which Qs describes the flowrate of oil following input Xi ; Pa describes the pressure developed in the cylinder due to Qs ; B is the bulk modulus of the oil; V is the average volume of oil under compression in the cylinder (V is more accurately expressed as V(O) + A · Xm , which allows for the expanding cylinder volume. However, this introduces the non-linear product Xm · Pa); and the other parameters were previously described.

Depending on the analyst the equations 3.1 to 3.4 could be differently arranged, or even different in content. The present equations are all linear, whereas several non-linearities can readily be included if the analyst considers them to be important to his response predictions.

3.2.2 Transfer Function

This is the most common form of model. A transfer function is a linear differential equation with constant coefficients, in which a chosen response variable is related to a specific input variable. Fig. 3.1 c shows the operational way such a differential equation is expressed, in which D denotes the operator d/dt . Simple cross-multiplication will yield the conventional differential equation. The transfer function can be derived by combining equations 3.1 to 3.4 so as to eliminate all variables other than input Xi and response Xm .

The transfer function is particularly descriptive, as it describes the time dependent ratio of response to input and is independent of input form, though not of the input variable chosen. Analysis and model solution procedures based on transfer functions have been highly developed (stability criteria, root locus, frequency response, system compensation, equation solutions, etc.). However, the necessity of linearization reduces their usefulness for hydraulic control system analyses. The coefficients can be awkward combinations of system parameters — see the definitions of K1 to K4 on Fig. 3.1 c . Such awkward arrangements are not helpful in the system design situation.

3.2.3 Vector-Matrix Models

Vector-matrix techniques allow concise representation of the dynamic relationships between a number of system variables and a number of simultaneous inputs. Fig. 3.1 d shows the vector-matrix model of the system under discussion. The chosen state variables are Xm , $\dot{X}m$ and Pa . Only one input, Xi , is included: however the model is readily extended to include other simultaneous inputs to the system —— possibly an external force Fi applied to the inertia in the line of motion. In essence, the system's state variables are related to the inputs experienced by inherent system parameters which make up the $\underline{A}$ and $\underline{B}$ matrix entries.

In effect, the vector-matrix model describes concisely a set of simultaneous transfer functions each of which relates a specific system response variable (state variable) to a specific input variable. While vector-matrix equations can include non-linear entries, most subsequent formal vector-matrix analyses apply only to linear situations.

3.2.4 Block Diagrams and Signal Flow Graphs

Both are diagrammatic representations of a system's dynamic equations. Figs. 3.1 e and f show the block diagram and the signal flow graph respectively for the system of Fig. 3.1 a . The diagrams can be formed from the system equations. However, many analysts prefer to form the block or signal flow diagram direct from the system schematic, thinking the system dynamics out as they construct the diagram. A transfer function or vector-matrix or equation set model is readily deduced from a completed block diagram or signal flow graph. Alternatively, the diagram itself is readily simulated for response solution. There is no particular advantage of block diagram relative to signal flow graph, or vice versa. The latter is a little more economical of linework, but the former bears a closer equivalence to the physical system. Both are capable of ready reduction (if linear) to a transfer function or vector matrix model, though this is hardly necessary or even desirable in this day of computers. Non-linear terms can be included on block diagrams and signal flow graphs, but care must then be taken in manipulating the diagram.

3.3 DERIVING THE MODEL

All approaches to modelling require insight of the physical laws and relationships which govern the behaviour of the various devices used in the system components, together with an appreciation of the interaction of the components and of the structure of the system. The successful formation of dynamic models is an art as much as a science, and experience and experiment are important assets.

The procedure for developing a model can be summarized:

1. Obtain a diagrammatic representation of the system to be modelled, showing clearly how the system is intended to function;

2. Select that response variable (or set of variables) study of which will provide clearest appreciation of system response;

3. Designate an input (or set of inputs) which will cause the system to respond in a realistic operational mode. It is not necessary to specify the form of the input at this stage, merely its physical variable and location;

4. List assumptions which are valid in the particular circumstances, and which will render modelling easier but still adequately valid. Typical assumptions which might be valid include: input drive speed constant; pressure relief valve does not lift; all driven inertias can be lumped in with the designated load inertia; all friction is viscous; pressure in the return lines is negligibly small (zero); motor-load coupling is rigid; the model is for small scale response predictions only; etc. This is an area requiring considerable thought for each modelling attempt. However, there is a lot of confirmed past experience available in the literature;

5. Imagine the system to be in equilibrium. This can be at zero value state, or at some preferred steady-state operational state. Note all initial values of system variables, if other than zero;

6. Imagine the designated input to be applied at the instant designated as $t = 0$;

7. Form equations describing the subsequent behaviour or actions of the various physical components which comprise the system.

The relationships so formed can be expressed in equation or signal flow diagram forms as previously discussed. It is necessary to ensure

- that correct relationships are used;
- that all relevant relationships are included.

It is desirable that dynamically irrelevant relationships be omitted in order to have the simplest model adequate for the particular analysis to be made.

The equations should be algebraic rather than numerical. The substitution of numbers for algebraic representation of parameters is best left until the model has been developed and is ready for solution.

3.4 SELECTION OF RELATIONSHIPS

For a particular model, some of the relationships to be used will be clear cut. For example, the origins of equations 3.1 (lever ratios) and 3.4 (Newton's law) should be obvious. A few relationships might be more ambiguous, requiring judgement by the analyst. Equation 3.2 is in this category. Equation 3.2 is a well known linear expansion of the physically obvious expression for flow through the valve,

$$Q = f\left(Xv, (Ps - Pa)\right) \qquad 3.5$$

which is often good for small-scale response (perturbation) analyses. However, if large-scale response (large Xm) is required, the analyst may prefer the well known orifice flow form,

$$Q = Kv \cdot Xv \cdot (Ps - Pa)^{\frac{1}{2}} \qquad 3.6$$

where Kv is a complex flow coefficient.

Equation 3.3 also has a degree of ambiguity in the compressibility term $V/B \cdot \dot{P}a$. An average value may be taken for V (volume of oil under compression Pa) if only small-scale motion of the mass is envisaged. If the motion is of large-scale, V should account for the expanding cylinder volume due to motion by using,

$$V = V(0) + A \cdot Xm \qquad 3.7$$

The aspect of ambiguous equation selection for dynamic analysis is common to all modelling procedures. It will be discussed more fully in Chapter 7.

3.5 SELECTION OF PARAMETER VALUES

Before a solution (simulation) can be made, the various coefficients appearing in the model must have numbers assigned to them. For a particular system, some of the numbers are readily assigned: for the example of Fig. 3.1, $L1$, $L2$, A , V , and Ck should be in this category. However, assignment of some numbers can be ambiguous or difficult. In the present case, valve flow coefficients $Kv1$ and $Kv2$, or Kv , and friction coefficient Rf are likely to be difficult to pin down, whilst bulk modulus B may be ambiguous —— anywhere in the range 500 to 1750 MPa .

This aspect of dynamic analysis also is common to all modelling procedures and will be discussed more fully in Chapter 7.

3.6 ANOTHER EXAMPLE

Fig. 3.2 a illustrates those components of a hydrostatic drive system which a particular analyst thinks are significant to the dynamic response of the system. He wishes to predict the load's speed change $\omega j(t)$ due to command changes $\phi i(t)$ to the pump swashplate angle. In forming Fig. 3.2 a , the analyst has

- decided to regard the input drive speed to the pump (ωp) , as constant (i.e. as a source). This eliminates primary drive (electric motor and shaft to pump) dynamics, and the inertia of the pump rotating assembly, from the analysis;

- decided to lump the internal leakages (case drains) of pump and motor and account for them via the leakage coefficient $K\ell$;

- decided that the resistance of the short hydraulic lines joining pump and motor do not significantly affect the response $\omega j(t)$; also that the return line pressure is so low relative to the drive pressure Pa that it can be neglected for the dynamic analysis;

- neglected torsional compliance of the shafting between motor and load;

- lumped the hydraulic motor and its shaft inertias in with the load inertia to give the total Ij ;

- combined motor and load friction, to be described by the viscous friction coefficient Rf .

Consider the system to be in equilibrium. When the swashplate angle is changed by ϕi radians, the volumetric displacement rate of the pump, Qp m^3/s , is described by,

$$Qp = Kq \cdot \omega p \cdot \phi i \qquad 3.8$$

where Kq is the volumetric displacement coefficient for the pump $m^3/rad/rad$; i.e. its swept volume per radian/sec angular velocity, per radian swashplate angle (m^3/s per rad/s per rad = $m^3/rad/rad = m^3$) .

The volume flowrate of oil effectively used at the hydraulic motor, Qa m^3/s , is given by,

$$\begin{aligned} Qa &= Qp - Q\ell - Qc \\ &= Qp - K\ell \cdot Pa - V/B \cdot \dot{P}a \qquad 3.9 \end{aligned}$$

where $Q\ell = K\ell \cdot Pa$ is the internal leakage flowrate,

a. Schematic for Dynamic Analysis

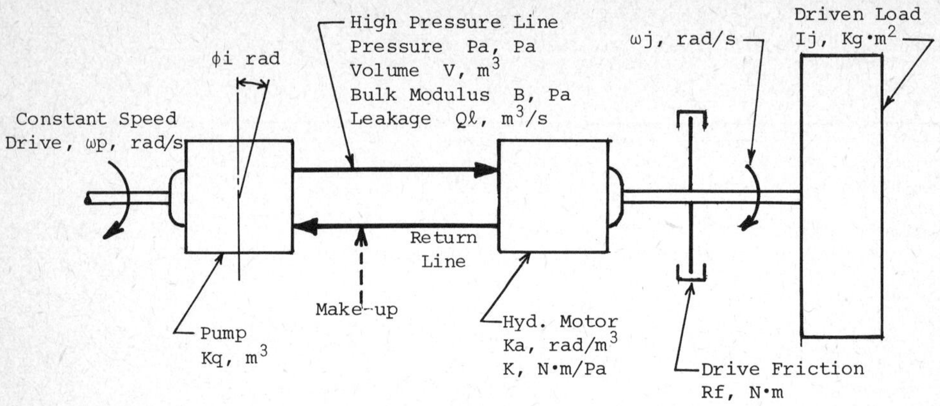

b. Block Diagram for System

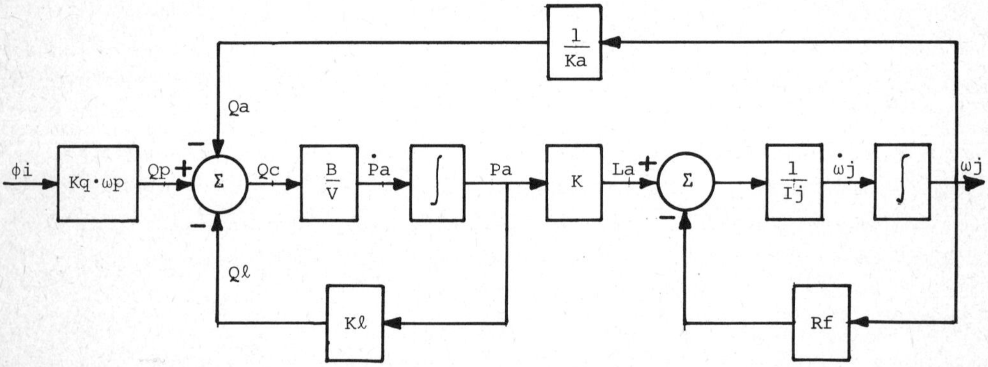

FIG. 3.2 HYDROSTATIC TRANSMISSION

$Qc = V/B \cdot \dot{P}a$ is the flowrate absorbed in compression of the oil,

$K\ell$ is leakage flow coefficient, $(m^3/s)/Pa$,

V is volume of oil under compression, m^3 ,

B is bulk modulus of oil, Pa .

Motor speed ωj due to Qa is given by

$$\omega j = Ka \cdot Qa \qquad 3.10$$

where Ka is the geometric flow speed coefficient for the motor, rad/m^3 ; i.e. rad/s per m^3/s .

The motor pressure-torque relationship is

$$La = K \cdot Pa \qquad 3.11$$

where K is the motor's pressure-torque coefficient, $N \cdot m/Pa$.

Applying Newton's law to the driven inertia,

$$La - Rf \cdot \omega j = Ij \cdot \dot{\omega} j \qquad 3.12$$

where $Rf \cdot \omega j$ describes the viscous frictional torque which resists motion.

The set of equations 3.8 to 3.12 is the model proposed for the system in order to study the dynamic response $\omega j(t)$ of the motor and load to swashplate angle changes $\phi i(t)$. The equations are in an informally structured form. They are readily combined to provide a transfer function of the form

$$\frac{\omega j}{\phi i} = \frac{K1}{K2 \cdot D^2 + K3 \cdot D + 1} \qquad 3.13$$

where the K's are composed of system coefficients, and D is operator d/dt . Alternatively, a block diagram is readily constructed, Fig. 3.2 b .

The equations are too simple to warrant expression as a vector-matrix model unless the analyst prefers to solve the model from its vector-matrix form.

If it is desired to study the system in a position control mode rather than the speed control mode, we simply add the equation

$$\theta j = \int \omega j \, dt + \theta j\,(0) \qquad 3.14$$

where θj is angular position of the load. This leads to the transfer function,

$$\frac{\theta i}{\phi i} = \frac{\omega j/D}{\phi i} = \frac{K1}{K2 \cdot D^3 + K3 \cdot D^2 + D} \qquad 3.15$$

and a corresponding change in the block diagram of Fig. 3.2 b . This formulation makes sense if swashplate angle ϕi is controlled from neutral, through a desired input cycle, and back to neutral, causing the load to move from one equilibrium position to another.

Note: The physically common notion of angular velocity in revolutions/minute or revolutions/second, can be used for pump and response speeds, ωp and ωj , if velocity coefficients Kq , Ka , Rf are suitably adapted. However to avoid errors in designation of numbers in models, the base SI unit radian/second should be used in all calculations and coefficients.

3.7 CONCLUSION

Conventional modelling procedures used in hydraulic control systems analyses have been reviewed briefly on the basis that the reader needs a reminder or orientation only. Some references which explain or apply these techniques more fully to hydraulic control systems follow.

3.8 REFERENCES

3.1 Blackburn, Reetof, and Shearer, Fluid Power Control, Tech. Press M.I.T., 1960.

3.2 Lewis and Stern, Design of Hydraulic Control Systems, McGraw-Hill, 1962.

3.3 Khaimovich, E., Hydraulic Control of Machine Tools, Pergamon, 1965.

3.4 Walters, R., Hydraulic and Electrohydraulic Servo Systems, Iliffe, 1967.

3.5 Merritt, H., Hydraulic Control Systems, Wiley, 1967.

3.6 Aizerman, M., Pneumatic and Hydraulic Control Systems, Pergamon, 1968.

3.7 Guillon, M., Hydraulic Servo Systems, Butterworth, 1969.

3.8 Kellar, G.R., Hydraulic System Analysis, Indust. Pub. Co., 1970.

3.9 McCloy, D. and Martin, H., Control of Fluid Power, Longman, 1973.

3.10 Prokes, J., Hydraulic Mechanisms in Automation, Elsevier, 1977.

CHAPTER 4

POWER FLOW MODELLING

4.1 INTRODUCTION

The intention here is:

- to discuss the case for modelling hydraulic control systems via study of the instant (dynamic) power flows taking place between the components of the systems;
- to introduce and demonstrate the power port description of hydraulic components and systems;
- to discuss the mathematical description of dynamic power flows between components, and in the system;
- to demonstrate the formation of a power port model for a particular hydraulic system.

Those readers familiar with these concepts may prefer to pass over or scan this Chapter and move directly into power bond graph formulation. Those for whom power flow modelling is relatively unknown should study this Chapter as an introduction to the more sophisticated power bond graph material.

Generally, hydraulic control systems are contemplated only where substantial amounts of energy have to be utilized to do work; that is, where large forces or torques have to be developed to move resisting loads. The amount of work done on the resisting load is dictated by the amount of energy utilized effectively. The rate at which the driven load moves is dictated by *the rate* at which it absorbs

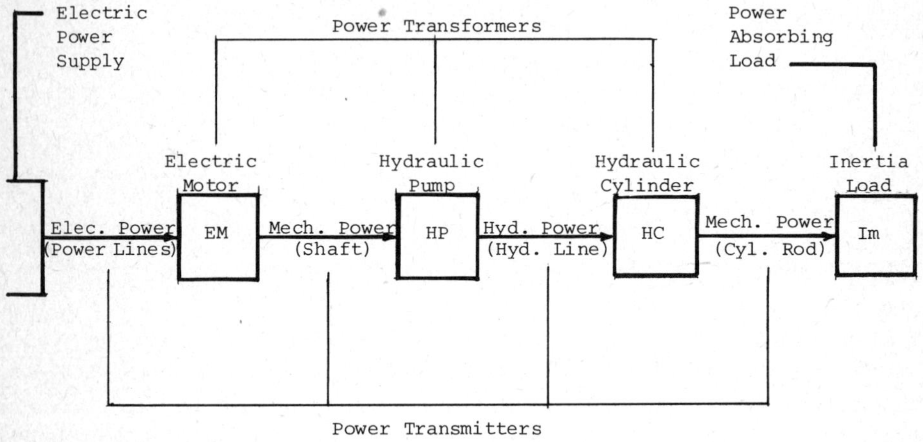

FIG. 4.1 POWER CHAIN FOR HYDRAULIC CONTROL SYSTEM

energy. A low energy-utilization rate means a slow motion imparted to the load; a higher energy-utilization rate causes faster motion. Thus, the dynamic response of a hydraulic control system is dictated by the dynamic energy flows (i.e. energy flow-rates) taking place.

In engineering, flowrate of energy is called power. That is, so many joules/second (watts) describes power. Power is a word with many facets, as a glance at a dictionary will show. One widely accepted and descriptive concept is that power flows. An electric motor draws energy at a required rate from the electric power supply mains. That is power flows from the mains to the motor and subsequently to the load being driven by the motor. A hydraulic pump can draw power from its prime mover (electric motor, diesel engine, aircraft engine, etc.) and transform it to hydraulic power capable of being utilized. Fig. 4.1 illustrates a simple hydraulic control system in which the electric motor, the hydraulic pump, and the hydraulic power cylinder, act as intermediaries between the source of power (the electric mains) and the load which requires moving. When the system is activated, power *flows* from the mains through the components and to the load. In this sense the term power flow is used to describe energy flowrate. Power flow will be static if its associated energy flowrate is constant. Power flow will be dynamic if its energy flowrate is time dependent.

Accepting the idea of power flow, it follows that system dynamic response is dictated by dynamic power flow. From this arises the concept that dynamic response should be predicted through study of dynamic power flow in the system. This idea is reinforced by realization that a hydraulic system *does* function via dynamic power interchange between its hardware components. The need then becomes to be able to describe the instant power flows taking place between the assembly of components which together comprise the system. The components of the system of Fig. 4.1 can be considered to be a power source (the mains supply); power transformers (the electric motor which accepts electric power and puts out mechanical power, the hydraulic pump which transforms mechanical input power to hydraulic power, the power cylinder which accepts hydraulic power and puts out mechanical power to the load); power transmitters (the shafting connecting pump to electric motor, the hydraulic line connecting cylinder to pump); and a power absorber (the driven load).

Further, hydraulic control systems are controlled by very low powered manual or automatic command signals, and can consequently be regarded as power amplifying in addition to power transforming and power transmitting systems.

Let us accept the natural prima facie case for investigating the dynamic response of hydraulic control systems through study of the power flows which take place following a command input to the system. The benefits of doing this will emerge in due course. Meanwhile, it is necessary to develop the idea further by considering the nature of the power flows. The basic requirement of dynamic analysis remains that a mathematical model of a proposed system be formed and then simulated (solved) to provide the required response information. Thus, we propose to seek a dynamic mathematical model through consideration of dynamic power flows.

4.2 POWER PORTS

Each component of a hydraulic control system has only a limited number of ways through which power can be taken in or given out. Neglecting all losses for the moment, an electric motor can take in power from the mains supply, and can give it out to its output shaft. That is, it has only two physical power flow connection points. These are known as power ports, or simply as ports. Fig. 4.2a illustrates the motor as a schematic, and as a block having two power ports (column 3).

A hydraulic accumulator has only one port through which it can accept or give out power. It is represented in Fig. 4.2b as a schematic, and as a block having a single power port.

A hydraulic pump can also be regarded as essentially a 2-port device, the ports being the drive shaft and the discharge port. However, the internal leakage (case drain) of a pump affects its power output and consequently the dynamic response of the system. It should be accounted for in the power port structure as a third port, through which power is lost, as illustrated on Fig. 4.2c, column 3.

The 4-way control valve is a 4-port device, with power able to flow into and out from it via the supply port, the two control ports, and the exhaust port. Fig. 4.2d illustrates this.

A hydraulic line is an obvious 2-port component, with power in or out only at its two ends (Fig. 4.2e). A shaft is similarly a 2-port device (Fig. 4.2f).

A double-acting hydraulic cylinder is functionally a 3-port device, as illustrated on Fig. 4.2g. It receives hydraulic power at one port, returns some of it to tank through a second hydraulic port, and gives out power through its piston rod.

A well designed and well placed system reservoir (tank) rarely affects a system's dynamic response. However, it receives power from the pump leakage, and from the return line from the control valve. A reservoir can be shown as a 2-port device receiving these two power flows (Fig. 4.2h).

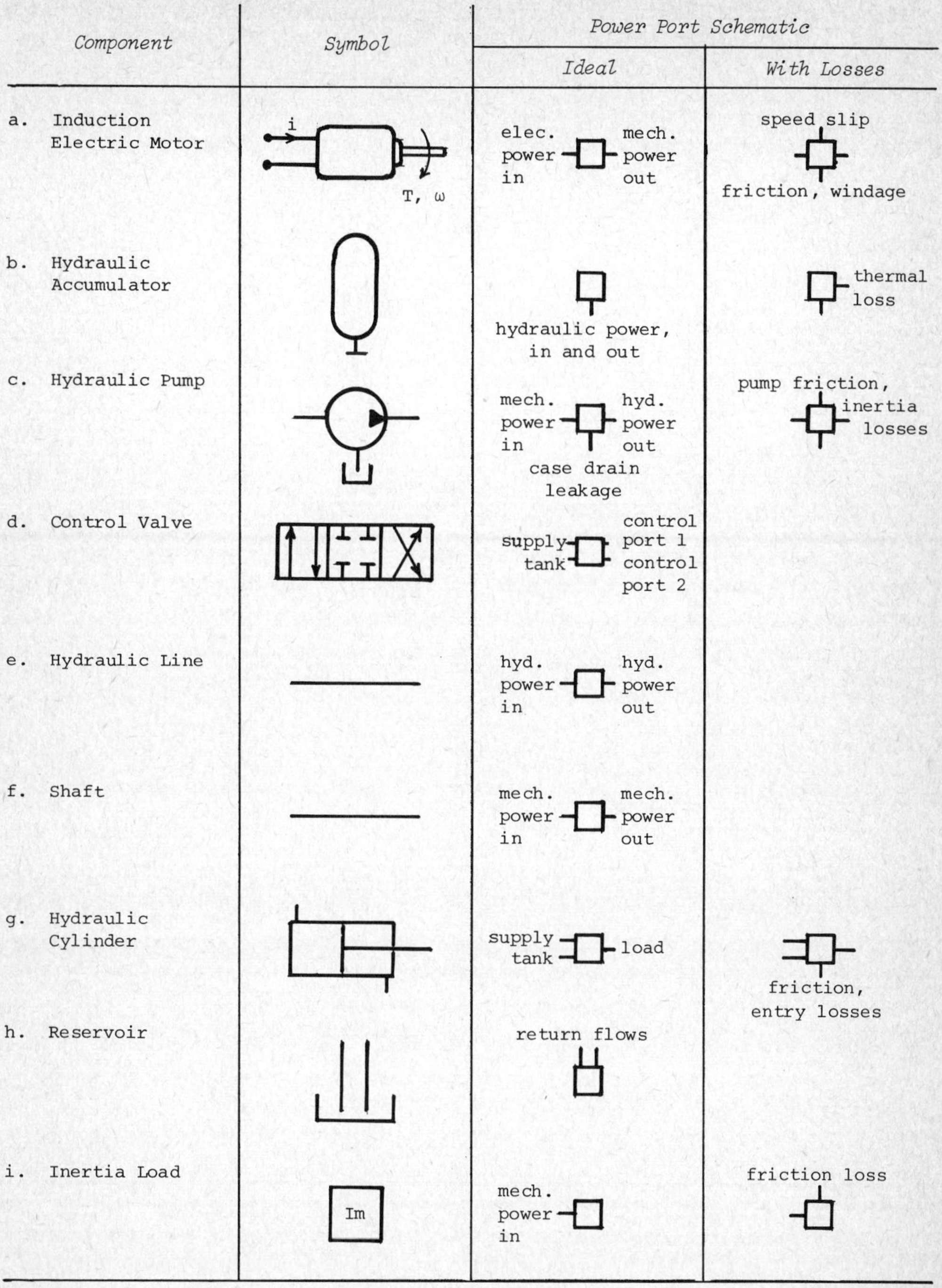

FIG. 4.2 SOME POWER PORT CONFIGURATIONS

A driven load of the kind illustrated on Fig. 4.1 is ideally a 1-port component, this being the mechanical connection whereby the actuator is connected to the load (Fig. 4.2i).

Thus, each component of a hydraulic control system can be regarded as an n-port device, with n usually between 1 and 4. Significantly, each designated port is an obvious physical reality of the component.

Basic power port designations are readily expanded to include loss and other effects which are thought, in a particular case, to affect a system's dynamic response. For example, an electric induction motor loses power to speed slip and should be regarded as a 3-port component. It may also lose significant power to internal friction and windage and could then be regarded as a 4-port component (Fig. 4.2a, column 4). Similarly, the n-port designation of a hydraulic pump could include an environmental port allowing for friction loss (Fig. 4.2c, column 4). Fig. 4.2, column 4 includes some power loss allowances for other devices. The designer-analyst should not feel compelled to use the most complete power port representation. Rather, he should include only those power flows which he feels will significantly affect the dynamic response of his particular system. The simplest model adequate for the purpose in hand should always be sought.

Power port blocks can be used to form a power flow schematic of a system. Simply couple the blocks as adjacent components are themselves coupled. Fig. 4.3 shows a power flow schematic for the system of Fig. 4.1, expanded to include a control valve and the reservoir, both of which manipulate or receive power in this system. A power port schematic provides a structure from which the set of equations which is the model of the system can be prepared.

In summary a power port schematic for a hydraulic control system is readily formed by:

- considering the power flowing into and from each component to be used in the system, and forming a set of power port blocks for the components. These component blocks are quite reuseable;

- simple joining of the component blocks to form the system power flow schematic. Such joining is directly analogous to physical connection of the hardware components to form the system. For example, the power output end of the electric motor power port is made one with the power input end of the hydraulic pump block. The line joining the two blocks is precisely analogous to the shaft which connects the two physical components. In both cases the connection is a power bond.

a. Detailed Schematic

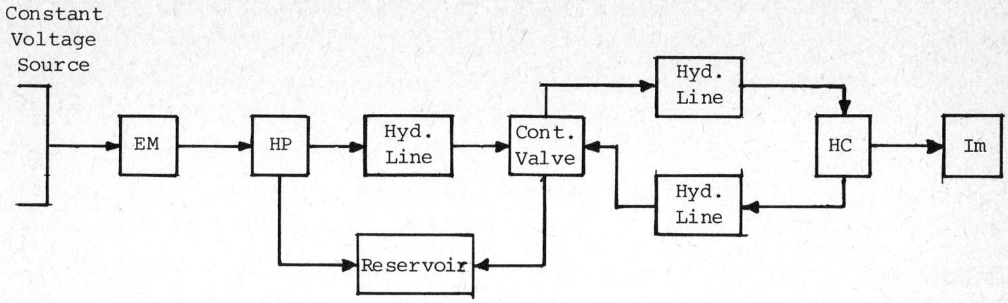

b. Shaft Compliance Recognized

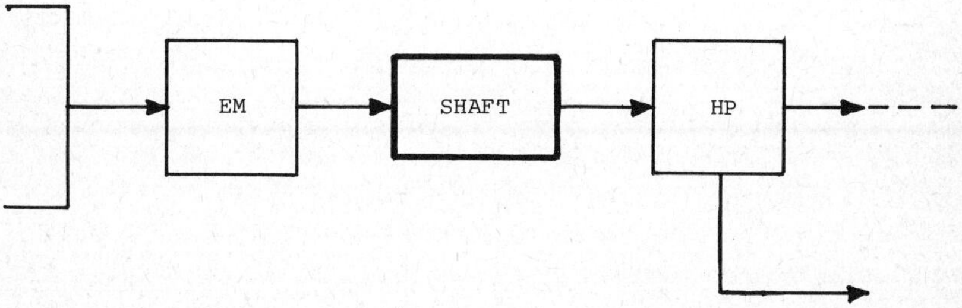

c. Pump Speed Assumed Constant

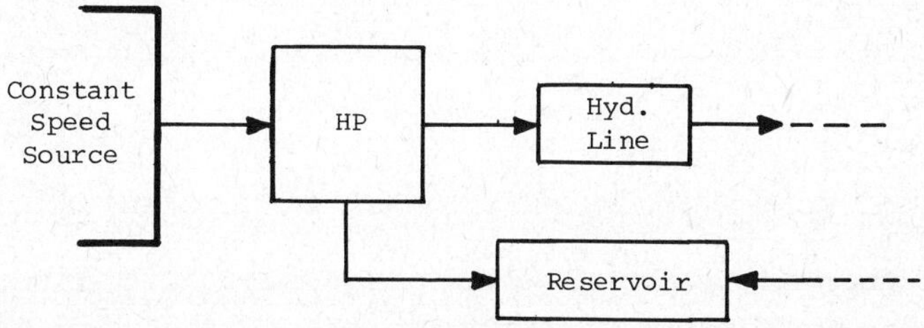

d. Constant Pressure Supply to Control Valve

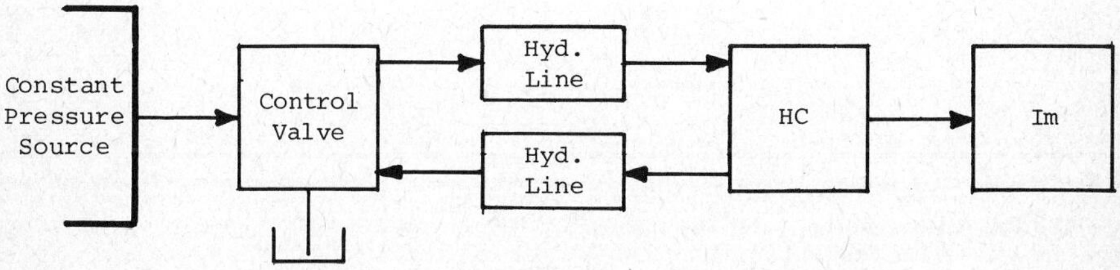

FIG. 4.3 POWER PORT SCHEMATICS

The power port schematic is readily expanded to allow for effects thought to affect system response. For example, if it is required to include the compliance of the shaft between electric motor and pump, place a block labelled SHAFT in the line joining the motor and pump blocks (Fig. 4.3b).

Also, a proposed schematic can be simplified to neglect an aspect considered to be insignificant for a particular response analysis. For example, the dynamics of the electric motor may have a negligible effect on the response being studied. In this case the motor is regarded as a source of constant speed, and its block is removed from the schematic (Fig. 4.3c). In essence, the source of constant voltage (mains supply) is replaced by a source of constant speed (but not constant torque!). Fig. 4.3d shows a further simplification, justified in some analyses, in which the pump dynamics also are neglected, making the pumping system a source of constant pressure. In each of these source simplifications it is not implied that power is a source, (i.e. independent); rather, that a particular variable is a source (i.e. independent). The associated power variable can vary. Finally, one or more of the system's hydraulic lines can be removed from consideration in the model simply by removing its block from the power port schematic.

Development of the model from a power port schematic will be demonstrated in Section 4.4.

4.3 DESCRIBING POWER FLOW

Power flow magnitude is described by the product of two simultaneous variables. For example,

. Electrical power flow	is	Voltage drop	•	Current
. Mechanical power flow	is	Net force (or torque)	•	Velocity (or angular velocity)
. Hydraulic power flow	is	Pressure drop	•	Volumetric flowrate
. Heat transfer power flow	is	Temperature drop	•	Heat conductance

Note that

- the first variable of each product is a potential variable; i.e. it offers the potential to induce power flow,
- the second variable of each product is a flowrate variable (current is coulomb/second or ampere, velocity is metre/second or radian/second, etc.).

Fig. 4.4a illustrates a simple hydraulic control system; Fig. 4.4b includes the power flow product variables, with the potential variables above and the flowrate variables below the component symbols. Arrows indicating the natural flow direction of power are included — power is drawn from the electric mains supply (source of voltage) and is delivered to the driven load. Neglecting all losses, power is conserved and

Voltage · Current = Force · Velocity

The relationship between the potential and the flowrate variables of a particular power product is of a cause-effect form. For example, force is the cause which induces velocity of the load mass. Voltage applied to an electric motor is a cause, and the resulting current drawn from the supply is an effect induced by the load impedance seen by the motor.

At first sight it might appear that the potential variable is always cause and the flowrate variable is always effect. This is not the case. It is not always obvious which variable of a power product is cause, and which is effect. Appreciation of cause-effect is a great aid to the understanding of a system's dynamic behaviour. It will be demonstrated in later Chapters that assignment of causality (i.e. decision on which variable of a power product is cause and which is effect) can be a key link in obtaining from a power flow diagram the correct set of dynamic equations for simulation.

Consider the system of Fig. 4.4a and b. It is proposed to decide which variable in each power product is cause and which is effect. We shall rearrange Fig. 4.4b to show cause variables above and effect variables below, rather than the Fig. 4.4b arrangement of potential variables above and flowrate variables below. Further, we shall designate the cause variable by an arrow in the direction of desired power flow, and the effect variable by an arrow directed against the power flow. This is a convention. However, it has physical significance, as the effect variables are *back effects* dictated by the dynamic nature of the system. For example, applied voltage is a source or constant and can only be designated as a cause as its magnitude is not affected by the load effects of the system. Designation of voltage as cause automatically requires current to be designated as an effect. However, this is also physically the case. The current drawn by the electric motor depends on the load applied to the motor *by the system*, i.e. the current *is a* back effect!

As discussed earlier, force applied to the load is readily perceived as a cause, and the subsequent velocity of the load is a consequent effect. There remains some ambiguity on this interpretation which bond graphs in later Chapters will clarify.

This leaves cause-effect decisions to be made for the hydraulic pump and for the

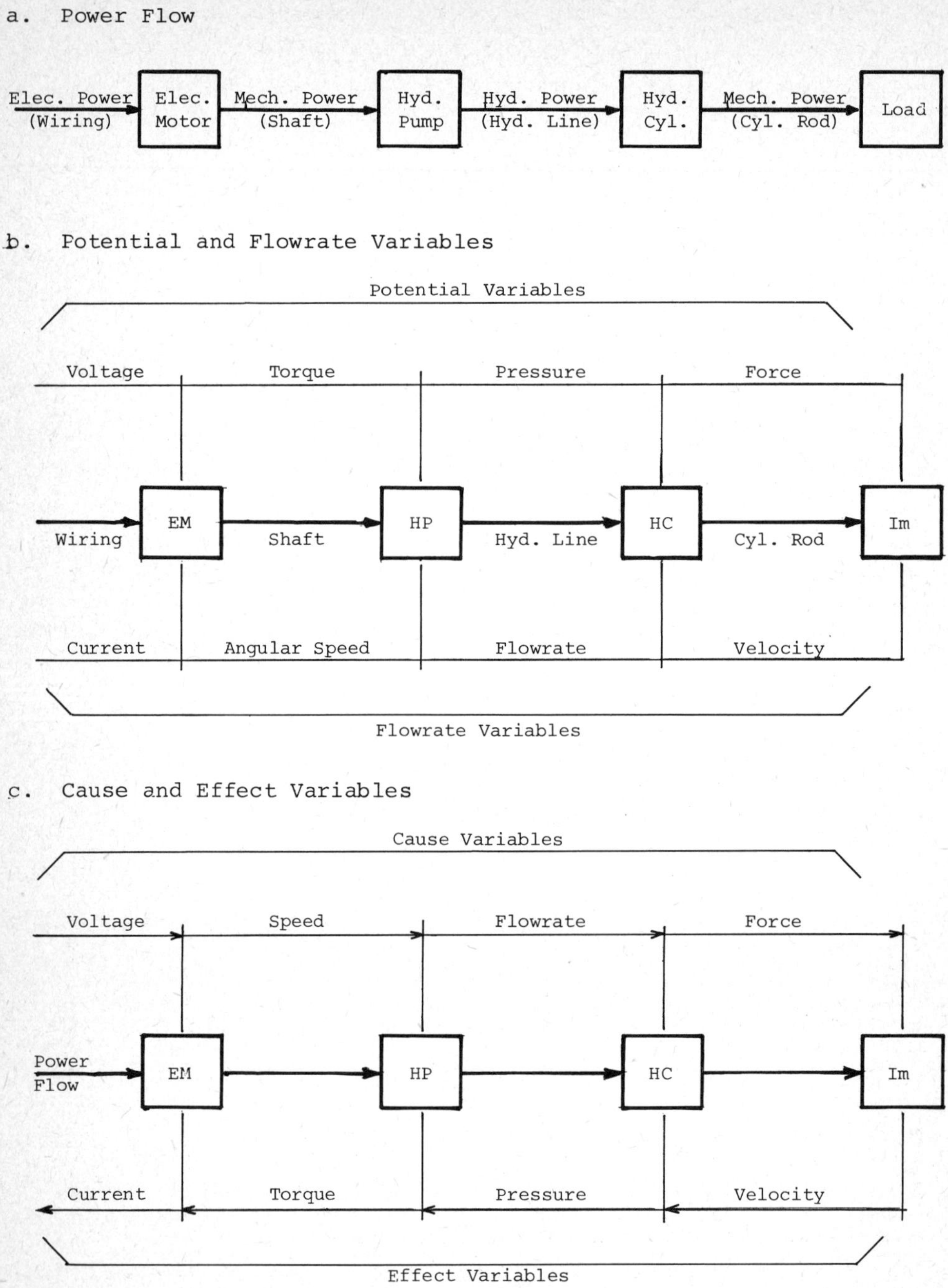

FIG. 4.4 POWER FLOW AND ITS VARIABLES

hydraulic cylinder. It is not necessary to consider components in the order in which they appear in the system —— doing this could make the task harder. Let us simply think about the action of the components, and decide the more obvious causalities. The remaining causalities will flow from these decisions. The system of Fig. 4.4 is not a constant pressure system. The pressure developed by the pump is load dependent. Thus, pressure is a back effect, and can be so designated on Fig. 4.4c, making flowrate a cause. The torque required at the pump shaft is also load dependent (pressure dependent if you like). Consequently, torque is also a back effect —— an effect variable —— making motor pump speed a cause variable. In this manner we complete the causality assignment for the system.

Check to see that physical sense is not violated. Voltage is applied to the electric motor, whose motion is resisted by the system load; this requires current to be drawn from the supply main. Resultant speed of the motor is transmitted directly to the hydraulic pump, which, being of positive displacement, causes positive flowrate of liquid to the cylinder. Liquid pumped positively into the hydraulic cylinder, whose piston motion is resisted by the inertia load, creates rapidly increasing pressure. This pressure acts on the piston to generate a force, which acts on the inertia load. The inertia load is put into motion. The magnitude of the inertia load dictates its acceleration and consequently its velocity, which is also the velocity of the piston. Piston velocity affects the rate of pressure change. Pressure dictates the torque required to drive the pump and thus the torque output required of the electric motor. The torque demanded of the motor dictates the current drawn by it. These actions do not take place in complete time sequence as implied here. They are occurring simultaneously, but are out of phase in the time domain. That is, it takes time for an action to fully develop following an applied cause. Predicting the sum total of these time effects is what dynamic analysis is all about.

Note that the effect variable does not necessarily result directly from its own cause variable. In the present example, this correlation occurs only for the inertia load, for which velocity is the effect of the causing force. Certainly current is not caused directly by the applied voltage; it is dictated by the shaft torque demand.

It cannot be assumed that the causality of a particular power variable will be the same in all systems and analyses. For example, it is sometimes appropriate to assume pressure a cause, and flowrate an effect. In constant pressure supply systems this will be the case. Making causal sense in a particular system is a great aid towards understanding the dynamic nature of the system, and in development of its dynamic model.

4.4 GETTING EQUATIONS

So far, we have reached the stage where a power flow schematic can be proposed. Thinking this out and making the associated decisions provides insight into those aspects of its components which will affect a system's dynamic performance. The ultimate objective remains the production of a set of equations (the dynamic model) which can be simulated (solved) to obtain the predicted behaviour of the significant system variables. It will become apparent that a power flow schematic can provide the structure from which a complete and suitably arranged set of equations can be written.

It is necessary to write expressions relating the variables and coefficients associated with each power bond (the line joining connected power ports). The relationships available are those utilized in all dynamic modelling procedures —— i.e. the basic physical laws and relationships describing the various actions which take place when a system is activated.

For example, consider the simple situation of Fig. 4.5a, for which supply pressure Ps is regarded as constant (a source). When the variable position 4-way closed-centre control valve is actuated by the displacement input signal Xv , the system will respond dynamically to move the inertia load. Fig. 4.5b shows the initial power port structure proposed to describe dynamic power flow following application of Xv . Hydraulic line resistance has been neglected. If power loss in the return line from cylinder through valve to tank is considered to be negligible as far as load dynamic response is concerned, the power flow schematic can be simplified to that of Fig. 4.5c. Each power bond (line) on Fig. 4.5c is allocated its power product variables, as follows:

$Ps \cdot Qs$ describes power flowing from source to control valve,

$Pa \cdot Qa$ describes power flowing from control valve to cylinder,

$Fm \cdot \dot{X}m$ describes power absorbed by the load (i.e. flowing from cylinder to load.

where Ps is the supply pressure, Qs the supply flowrate;

Pa is the pressure developed in the cylinder, Qa the effective flowrate to the cylinder —— i.e. the volumetric displacement rate of the piston in the cylinder;

Fm is the net force applied to the driven load, and $\dot{X}m$ the velocity developed by the load.

Also, we propose to account for a power loss due to friction associated with the

a. The System

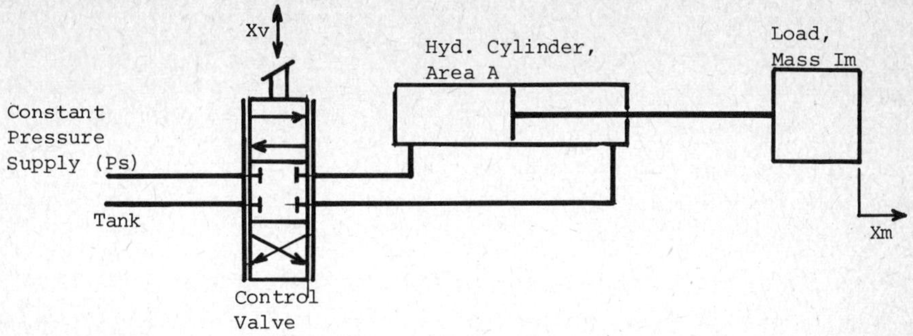

b. Initial Power Port Structure Proposed

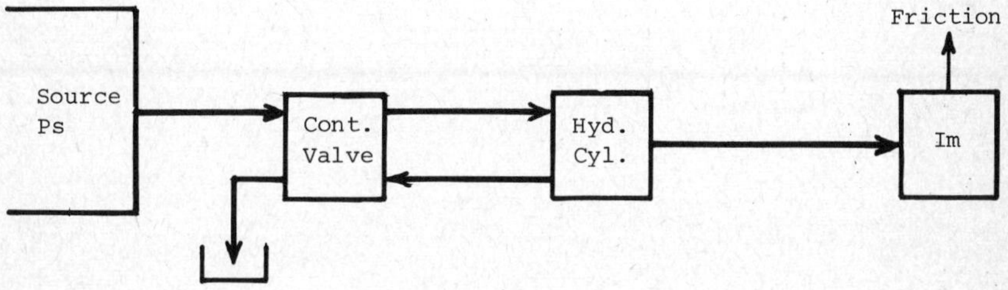

c. Simplified Power Port Structure, with Power Variables on Each Power Bond

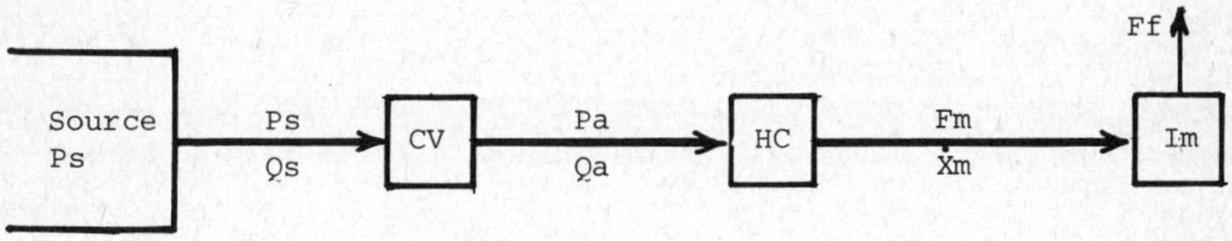

FIG. 4.5 POWER PORT STRUCTURE FOR SIMPLE HYDRAULIC SYSTEM

driven load, and which is described by $Ff \cdot \dot{X}m$, where Ff is the force required to overcome friction.

Being satisfied that the power flow schematic of Fig. 4.5c is adequate for our purpose, we now require to prepare from it an appropriate and complete set of equations. This requires an equation for each power variable, expressed in terms of only other system power variables and system coefficients. Further equations which might be required to describe variables other than these *power state variables* can be added if desired. Environmental power loss effects, such as load friction in the present case, can be described as separate equations, or can be built into the power state equations, whichever the analyst may prefer. For the present case, we can write the six power state equations:

$$Ps = \text{constant} \quad \text{(an assumption)} \qquad 4.1$$

$$Qs = f\left(Xv, (Ps - Pa)\right) \quad \text{(orifice flowrate, form to be decided)} \qquad 4.2$$

$$Pa = -B/V \cdot \int (Qs - Qa)\,dt + Pa(0) \quad \text{(capacitive effect)} \qquad 4.3$$

$$Qa = A \cdot \dot{X}m \quad \text{(piston swept volume rate)} \qquad 4.4$$

$$Fm = Pa \cdot A - Ff \quad \text{(net force on } Im \text{)} \qquad 4.5$$

$$\dot{X}m = 1/Im \cdot \int Fm\,dt + \dot{X}m(0) \quad \text{(Newton's law, integral form)} \qquad 4.6$$

We look now for any additional equations required

- to specify any significant response variables other than the power state variables,
- to specify any non-constant coefficients used in the state equations,
- to specify the controlling input.

In the present example, this leads to the auxiliary equation set:

$$Xm = \int \dot{X}m\,dt + Xm(0) \qquad 4.7$$

(Xm is a respone variable of primary interest)

$$V = V(0) + A \cdot Xm \qquad 4.8$$

(V is volume of oil under compression)

$$Ff = \text{to be decided} \qquad 4.9$$

(friction force)

Xv = to be specified (control valve displacement input) 4.10

Coefficients in the equations not already identified are:

A - effective area of hydraulic ram

B - bulk modulus of the fluid in the system

V - volume of fluid under compression Pa (i.e. volume in hydraulic line from valve to cylinder plus the cylinder chamber). V will vary with piston displacement and can be expressed $V = V(0) + A \cdot Xm$.

These equations will be readily recognized by those familiar with dynamic analysis of hydraulic systems. The precise relationship of equation 4.2 can be chosen from several widely used forms, including:

$$Qs = Kv1 \cdot Xv - Kv2 \cdot Pa \qquad 4.2a$$

a linear equation with Kv1 and Kv2 as flow and pressure coefficients respectively; or,

$$Qs = Kv \cdot Xv \cdot \mathrm{sgn}\,(Ps - Pa) \cdot (Ps - Pa)^{\frac{1}{2}} \qquad 4.2b$$

a non-linear flow description with Kv an orifice coefficient.

Similarly, the analyst can choose the precise friction force expression from forms such as

$$Ff = Kf \cdot \dot{X}m \qquad 4.9a$$

which describes linear viscous friction of coefficient Kf ; or,

$$Ff = \mathrm{sgn}\,\dot{X}m \cdot Fu + Kf \cdot \dot{X}m \qquad 4.9b$$

which includes a coulomb friction term Fu .

It is quite an advantage to be able to defer decisions on the forms of some of the more ambiguous relationships to be used in the final simulation.

The set of equations 4.1 to 4.10 is the dynamic model proposed for the system of Fig. 4.5. The model is comprised of a set of quite simple basic relationships in algebraic and integral form. Note that $\dot{X}m$ is a state variable, and thus its appearance in an equation does not make that equation of differential form —— it is not intended that $\dot{X}m$ be derived from Xm .

Actually, the selection and arrangement of equations is not as clear cut as appears here. There are some causal decisions to be made in order to produce integral rather than differential expressions. For computer simulation it is far preferable to have equations in integral rather than derivative forms. In the present case, equations 4.3 and 4.6 reflect such causal decisions, and these affect the arrangement of the other equations. A power flow variable should be the object of one equation only, and each power flow variable is to be the object of an equation.

A more substantial system or analysis would result in a much larger number of equations being required, and considerable confusion on the equation arrangement, and even on completeness or otherwise of the equation set, can arise. Unruh [4.1] and Young [4.2] have separately developed procedures for formally sorting and arranging equations from power port structures to satisfy both causality and completeness. Selecting of equation forms, and the obtaining of numerical values for coefficients are discussed more fully in Chapter 7.

4.5 AN EXAMPLE

The following example is adapted from Reference 4.3.

Fig. 4.6a illustrates a hydrostatic drive system whose dynamic response and performance characteristics are required prior to purchasing the components and constructing the system. Fig. 4.6b shows a power port diagram prepared for the system. Each line between blocks is a power bond, representing a physical power-transmitting connection. Each line has two power state variables associated with it, their product describing the power flowing. At first sight it appears that there will be 2n power state variables, where n is the number of power bonds present. However some bonds share a power state variable. For example, Pp is the pressure on the bond between pump and relief valve, and also on the bond between relief valve and filter. Also, ωa is the velocity (flowrate variable) on both the hydraulic motor-load bond and the load friction power dissipation bond. Examination of the power port diagram reveals the following set of power state variables, starting from the assumptions that the electric motor runs at constant speed, and that the charge pump power requirement does not affect the dynamics of the main system:

Sω	constant speed of electric motor, and hydraulic pump
T	torque developed in the motor-pump shaft
Pp	pressure developed by the pump
Qp	flowrate developed by the pump
Pe	reservoir pressure, often assumed zero
Qs	flowrate from reservoir to pump
Qℓp	leakage flowrate from pump
Qr	flowrate through pressure relief valve to reservoir
Qb	flowrate into filter (also from filter)
Pb	pressure at discharge side of filter

a. Circuit Diagram

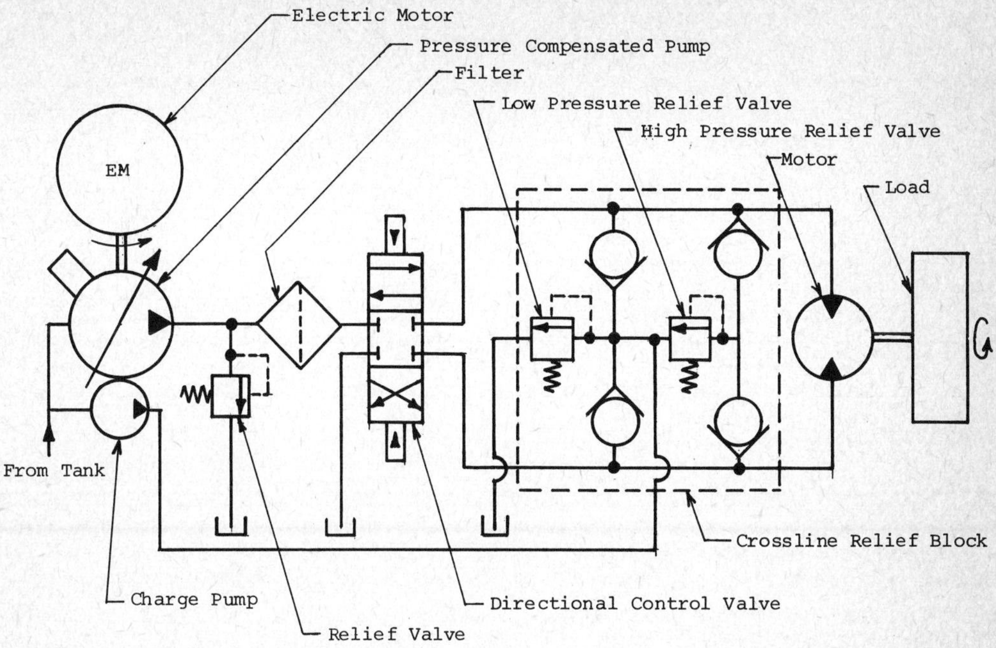

b. Power Port

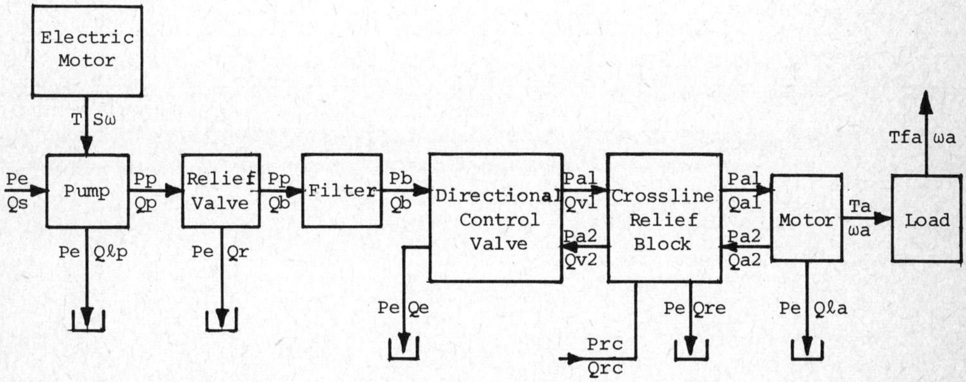

FIG. 4.6 A HYDROSTATIC DRIVE AND ITS POWER PORT DIAGRAM

Pa1	pressure developed in line between control valve and hydraulic motor
Qv1	flowrate from control valve towards motor
Qa1	flowrate received by hydraulic motor
Pa2	pressure developed in return line between motor and control valve
Qa2	flowrate from motor towards control valve
Qv2	return flowrate entering control valve
Qe	discharge flowrate from control valve to reservoir
Qre	flowrate from crossline relief block to reservoir
Prc	pressure in charge line to crossline relief block
Qrc	flowrate in charge line to crossline relief block
Qℓa	leakage flowrate from hydraulic motor
Ta	torque developed in motor-load shaft
ωa	angular velocity of motor and load
Tfa	friction torque

To form the model, an equation for each of these 24 power state variables is required. If numbers can be provided for the various coefficients, the set of equations can be simulated to provide the sought-after dynamic response information. The system can be controlled or disturbed by adjusting the state of the variable delivery pump, by adjusting the 4-way control valve, or by adjusting load inertia or friction. This example and its equations are developed more fully via bond graphs in Chapter 8.

4.6 SUMMARY

- The essential components of a hydraulic control system can be placed in one of three groupings:

  - power transforming components (electric motors, pump, actuators, etc.),

  - power transmitting components (electric supply wiring, shafts, hydraulic lines, cylinder rods, etc.),

  - power control components (pressure and flow control valves, etc.).

- The dynamic response of a hydraulic control system is dictated by the dynamic power interchanges taking place between the components of the system.

- Power can enter or leave a component through only a small number of readily identified and physically meaningful power ports. Adjacent components are connected together functionally only via these power ports.

- A dynamic model of a system, expressed as a set of simultaneous action equations, can be developed from power flow considerations. A power port block diagram is first formed to provide the structure from which the equation set will be prepared. Each line (power bond) joining power port blocks is used for power interchange.

- Power flow requires description as the product of two variables.

- Each power variable is a power state variable which is the object of one of the set of power state equations which comprise the basic dynamic model of the system.

- Using preferred causality, the power state equations can be arranged in algebraic and integral forms suitable for computer solution or simulation.

- System power port structures are modular and closely resemble the modular construction of the hydraulic system itself. This allows components to be readily changed, added, or deleted from a system's power port block diagram.

- Dynamic coupling effects are inherently included in models derived from power port structures.

- The advantages relative to conventional modelling procedures of the foregoing points are quite significant.

The development of power port models will not be pursued further, as the Power Bond Graph techniques to be developed in the next Chapters are a more advanced and formal development of power flow modelling.

4.7 REFERENCES

4.1 Unruh, D., A Standard Format for Mathematical Models of Fluid Power Systems, Proc. N.C.F.P., V26, 1972.

4.2 Young, M.Y., Digital Simulation of Hydraulic Control Systems, M.Eng.Sci. Thesis, Monash University, Australia, 1973.

4.3 Stecki, J.S. and Cuthbertson, S.C., The Effects of Changeover to Fire Resistant Fluids on Hydraulic Performance - Part 2: System Criteria, FRH Jnl., Vl, N2, pp 167-173, FPRC, Oklahoma State University, 1981.

CHAPTER 5

POWER BOND GRAPHS

5.1 INTRODUCTION

So far it has been proposed

- that prediction and subsequent optimization of dynamic response should be part of the procedure for the design of hydraulic control systems,
- that inadequate availability of dynamic models in forms suitable for use by the designer-analyst is the main obstacle,
- that conventional model forms are not particularly suitable for use in the system design situation,
- that dynamic power flows are particularly descriptive of the physical actions dictating the dynamic response of hydraulic control systems,
- that power flow models offer the system designer modular reuseable model forms which are close to direct representation of the similarly modular hydraulic system,
- that power flow modelling allows increased separation of the essential steps, and consequent increased formality, in forming a set of dynamic equations (the model) suitable for solution by computer.

The present Chapter introduces the *power bond graph* technique for developing dynamic models of powered control systems. It has become customary to refer to power bond graphs simply as bond graphs. Although the present author considers this to be an undesirable abbreviation, it is adopted for the Text. Bond graphs have been applied to a wide variety of dynamic situations, including vibration analyses, thermodynamic and fluid dynamic situations, and physiological systems. However, they are used in this Text solely as applicable to hydraulic control systems, for which they have a quite particular affinity. Only those aspects of bond graphs which have been found to be useful in the hydraulic control systems area will be discussed and utilized. This includes representation of electric motors, electric servovalves, inertia loads, and other such devices which occur within the field of hydraulic control systems. References 5.1 to 5.4 provide a more historical and far more complete information on bond graphs and their applications.

A bond graph is a dual-signal flow diagram formed from a limited set of symbols. From it a properly ordered and arranged set of dynamic equations can be prepared in a form suitable for computer solution. A new but very limited set of terms and symbols has to be learned. In effect, the multiport power flow approach of Chapter 4 is made more general, more concise, and more formal through the (power) bond graph symbols. We shall begin by introducing and discussing the basic terms and symbols which will be used in subsequent development of bond graph derived models. The reasons why bond graph techniques are being proposed in this Text as superior in the design-orientated analysis area will be presented, discussed, and demonstrated after the reader has gained an appreciation of what bond graphs are.

The strategy behind use of power bond graphs can be summarized:

- to use a very limited number of versatile general terms and symbols to provide a rational graphical structure which describes the presence, and the interactions, of effects which affect the dynamic performance of the system;
- to allow ready formation of, and subsequent changes in, the structure as is required in the creative design situation;
- to use the model structure to formally prepare a rational and adequately complete set of equations suitable for computer simulation of the system.

5.2 BOND GRAPH TERMS AND SYMBOLS

5.2.1 Effort and Flow Variables

In Chapter 4 it was illustrated that power flow is described by the product of a potential variable and an associated flowrate variable. Power bond graph researchers and users have accepted the terms *effort* and *flow* to mean potential and flowrate respectively. Thus, in a hydraulic line, pressure is the *effort* variable and volumetric flowrate is the *flow* variable. In this Text, E and Q will be used as the general symbols for effort and flow respectively. In specific situations, specific symbols will be used; for example P for effort (pressure) and Q for flow (flowrate) in hydraulic line power flow.

5.2.2 Sources

One or more of the power variables associated with a dynamic system can be constant or can be assumed to be constant for purposes of analysis. For example,

- the voltage applied to an electric motor is assumed to be constant even though the power drawn from the electricity supply by the motor fluctuates,
- some hydraulic control systems are analysed on the basis that supply to the control valve is at constant pressure,
- in most situations, pump drive speed may be regarded as constant.

In bond graph terms, each of these constants is called a *source*, and is designated by the symbol S . In general, in the present Text,

SE denotes an effort source

SQ denotes a flow source

although specific effort and flow variables will be specifically described. For example:

SP denotes a source of pressure

SQ denotes a source of volumetric flowrate.

Designation of a power variable as a constant does not imply that the associated power is constant. In fact, in a dynamic situation power will rarely be constant. The product variable associated with the designated source can vary, thus allowing power to vary. For the electric motor, voltage (effort) is constant but current (flow) varies as dictated by the load on the motor. For a constant supply pressure hydraulic system, flowrate can vary and power flow will vary accordingly. For power flow to be constant, either both effort and flow variables must be sources, or both must be allowed to vary such that their product remains constant. The former case is not compatible with dynamic states, though the latter may be sought in some control system designs.

A power state variable which is independent of the system, though not necessarily constant, is also regarded as a source.

5.2.3 Power Bonds

A path through which power flows is represented by a line known as a *power bond*.

Each power bond has an effort and a flow variable associated with it. In general it is represented

$$\frac{E}{Q}$$

or, in a particular case such as a hydraulic line,

$$\frac{P}{Q}$$

5.2.4 Power Transformers

As discussed in Chapter 4, many of the major components in a hydraulic control system are primarily power transforming devices. Thus, an electric motor converts electric power to mechanical power; a hydraulic pump converts mechanical power to hydraulic power; a hydraulic actuator converts hydraulic power back to mechanical power. Power transformation is represented in bond graphs as the symbol TF in the centre of a power bond, as follows:

$$\text{———— TF ————}$$

and,

$$\frac{E1}{Q1}\ \text{TF}\ \frac{E2}{Q2}$$

which means that

E1 transforms to E2
Q1 transforms to Q2

while power is conserved such that

$$E1 \cdot Q1 = E2 \cdot Q2$$

For the simplest transformations, E1 is related to E2 by a constant, and Q1 to Q2 by the inverse of the constant. In these cases the transforming modulus is noted below the TF symbol. For example, power transformation in a hydraulic cylinder is described by:

$$\frac{P}{Q}\ \underset{A}{\text{TF}}\ \frac{F}{\dot{X}}$$

where P and Q are the pressure and volumetric flowrate acting in the cylinder;
F and $\dot{X}$ are the force and velocity induced on the cylinder's piston and rod;
A is the effective cross-sectional area of the piston.

The symbol means simply

$$P \cdot A = F$$
$$Q/A = \dot{X} .$$

Note that power is conserved,

$$(P \cdot A) \cdot Q/A = P \cdot Q = F \cdot \dot{X} .$$

As another example, if a hydraulic pump is regarded as an ideal power transformer, it would be represented in a bond graph as,

$$\frac{T}{\omega}\; TF \;\frac{P}{Q}$$
$$Vp^{-1}$$

which means

$$T \cdot Vp^{-1} = P$$
$$\omega \cdot Vp = Q$$
$$T \cdot \omega = P \cdot Q$$

where T (torque) $\cdot$ ω (angular velocity) describes the power into the pump at its drive shaft,

P (pressure) $\cdot$ Q (flowrate) describes the power delivered by the pump if all losses are neglected;

Vp is the swept volume per radian of the pump.

It is enlightening to realize that in most hydraulic system pump situations, speed of the pump drive is intended to be constant, while torque developed is dependent on the pressure generated in the system. On this cause-effect basis, we might prefer to write the equations

$$Q = \omega \cdot Vp$$
$$T = P \cdot Vp$$

i.e. showing the effect variables as functions of the cause variables. Such causal decisions will be discussed shortly.

For a fixed displacement pump, Vp is constant. However, for a variable displacement pump, Vp is a function of the variable pump stroke. This would be recognized in bond graph terms as the *modulated transformer*,

$$\frac{T}{\omega}\ \text{MTF}\ \frac{P}{Q}$$
$$Vp^{-1}()$$

where () embraces whatever variable(s) control Vp .

Losses and dynamic effects which occur during power transformations are accounted for separately from the actual TF operation, which is treated as ideal; representation of losses and dynamic effects in bond graphs will be discussed and demonstrated shortly.

Gyration

A variation sometimes arises wherein the input effort variable transforms to a flow variable, and the input flow variable transforms to the output effort variable. This power transduction is called *gyration*, and its power bond graph symbol is GY , as follows

$$\frac{E1}{Q1}\ \underset{K}{\text{GY}}\ \frac{E2}{Q2}$$

the implication being that

$$E1 \cdot K = Q2$$
$$Q1/K = E2$$

rather than

$$E1 \cdot K = E2$$
$$Q1/K = Q2$$

which is the TF case.

The induction electric motor is an example. Applied voltage E (effort) induces speed ω (flow) of the motor field windings; the torque T induced requires current i to be drawn. In bond graph symbols this is described by

$$\frac{E}{i}\ \underset{K}{\text{GY}}\ \frac{T}{\omega}$$

which means

$$E \cdot K = \omega$$
$$i/K = T$$

with power conserved such that

$$(E \cdot K) \cdot i/K = E \cdot i = T \cdot \omega$$

Note: ω is <u>not</u> motor shaft speed, which will be less than ω by the amount of speed slip necessary to develop torque in the induction motor. Further development of the bond graph of an induction motor will be demonstrated shortly.

As with TF's , gyrators can be modulated in which case they would be designated as MGY with the appropriate modulating function indicated underneath.

5.2.5 Dynamic Effects

We come now to those effects which dictate the dynamic nature of a control system and which govern its dynamic response. These actions can be placed into one of the three general classifications,

- Resistive Power Dissipation
- Capacitive Power Storage
- Inertive (or Inductive) Power Storage

The power bond graph symbols for these are respectively

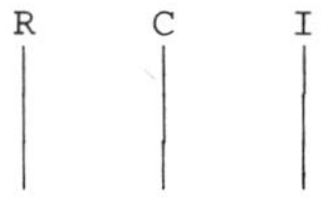

That is, a power bond leading to one of the three basic dynamic action classifications. Each classification will now be described through examples.

5.2.6 Resistive Power Dissipation

Any effect governed by an algebraic relationship between net effort and flowrate can be classified as a *resistive effect*. Resistive effects dissipate power. Resistive effects which can affect the dynamic response of hydraulic control systems include:

- friction in all of its forms. In particular, friction associated with driven loads is a significant factor in the dynamic response of hydraulic systems;

- pipe line pressure drops. Power flow in a pipeline is described by the product of pressure and flowrate. For a pipeline, the flowrates at entry and exit are essentially the same. However, there will be a pressure drop, and consequently a power loss, between entry and exit;

control orifices of various kinds (flow control valves, pressure relief valves, etc.). These too cause pressure drops at essentially identical inlet and exit flowrates and so dissipate power.

The examples all illustrate that power dissipation actions are resistive in nature. Further, they cause irreversible power loss. The bond graph symbol recognizing a power dissipation action is a power bond with the letter R (for Resistive effect), at its end. The R indicates that power transmission in the region of the bond is dictated or affected by a resistive effect. Power flow in the bond is described by the product of an effort variable and a flow variable. Thus the symbol becomes

R
E | Q

and its implication, to be developed later, is that a resistive effect R is present and that it describes a relationship

$$E = R(Q) \;, \quad \text{or} \quad Q = R(P) \qquad 5.1$$

where R() denotes "a resistive function of".

For example, consider the pressure drop in a hydraulic line. It can be recognized in bond graph terms as

Rh
ΔPh | Qh

If flow is assumed laminar, the linear resistance relationship can be utilized to provide

$$Qh = 1/Rh \cdot \Delta Ph \qquad 5.2$$

where Rh is the laminar resistance coefficient for the hydraulic line.

If, on the other hand, flow in the line was turbulent, the Rh effect cannot be accounted for via a simple coefficient. The pressure drop-flowrate relationship is better described by the non-linear equation

$$Qh = Kh \cdot \text{sgn}\, \Delta Ph \cdot \Delta Ph^{\frac{1}{2}} \qquad 5.3$$

where Qh is the volumetric flowrate in the hose;

ΔPh is the magnitude of the pressure drop across the hose;

Kh is a parameter describing hose geometry, fluid properties, and flow coefficients.

Irrespective of what Rh relationship is eventually chosen to relate Qh to ΔPh , the bond graph symbol remains as shown. The symbol recognizes the action, not the specific equation to be used. The symbol indicates that flowrate is a function of pressure drop, or vice versa. Equation selection comes later.

Friction can be included in bond graphs as a resistive effect Rf ; for example, as

which means

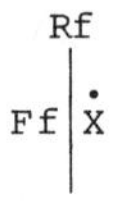

where Ff is friction force resisting motion $\dot{X}$.

The implication is that $Ff = Rf(\dot{X})$; that is that friction force is a resistive function of velocity. In the simplest case of viscous (linear) friction, Rf is a simple coefficient and the equation implied is

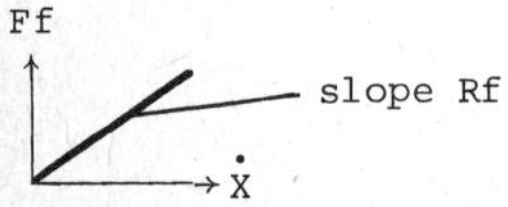

$$Ff = Rf \cdot \dot{X} \qquad 5.4$$

In the more general case in which coulomb friction (stiction) is also present, we may need to describe

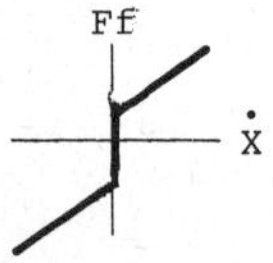

or

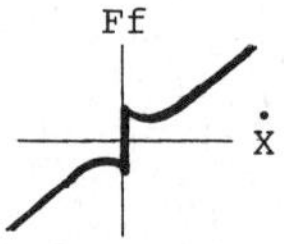

in the Rf function. This will be further developed in Chapter 7.

An action or device provides an R effect if it follows the concept that effort drop induces flow, or flow is accompanied by effort drop.

5.2.7 Capacitive Power Storage

A device provides a capacitive effect if net *flow* into it causes increased *effort* within it. Capacitance implies energy storage and thus power storage characteristics. Capacitive power storage devices include:

- mechanical springs,
- drive shafts in torsion,
- mechanical compliance in general,
- hydraulic accumulators,
- electric capacitors,
- in general, oil-containing volumes in hydraulic control systems (filters, hydraulic hoses, cylinder end volumes, etc.).

In general, these devices can store and give up potential energy in manners affecting dynamic performance of the system of which they are part. The bond graph symbol for a capacitive effect is:

the line (power bond) representing the path through which power can flow to or from the particular capacitive effect recognized by the C at the end of the bond. Implied on the bond are the effort and flow variables associated with the particular C being considered,

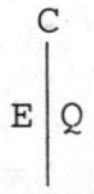

and the further implication is that

$$E = C(Q) \quad \text{or} \quad Q = C(E)$$

where C( ) denotes "a capacitive function of".

In the simplest case of linear capacitance, the function becomes

$$E = 1/C \cdot \int Q \, dt + E(0) \qquad 5.5$$

- i.e. for an electrical capacitor

$$E = 1/C \cdot \int i \, dt + E(0) \qquad 5.6$$

  (Faraday law, E is voltage, i is current)

- i.e. for a linear spring

$$Fk = 1/Ck \cdot \int \dot{X}\, dt + Fk(0) \qquad 5.7$$

where Fk is the force developed in the spring;

Ck is the compliance of the spring;

$\dot{X}$ is the velocity of one end of the spring relative to the other end.

Note that equation 5.7 is the dynamic form of the static spring relationship $Ff = 1/Ck \cdot X$

. i.e. for linear oil compressibility in a fixed chamber or volume

$$P = 1/C \cdot \int Q\, dt + P(0) \qquad 5.8$$

where P is the pressure in the chamber;

C is the hydraulic capacitance of the chamber; ideally, $C = V/B$ for a fixed volume, where V is the volume of oil under compression, and B is bulk modulus of the oil;

Q is the flowrate of oil into the chamber.

As with the R effects, there is no need for a designated C to be a coefficient. In a particular case, the C effect linking a potential (effort) to a flowrate (flow) may be a non-linear expression, or even empirical data. Several C relationships will be demonstrated later in the Text. Note that we have, in each case, expressed the potential variable as a function of the flowrate variable. This satisfies the basic causality requirement that the effect is a result of the cause.

The capacitive effect is referred to as "power storage". This does not mean that power is always flowing into a C element. In dynamic action, power can be flowing *from* the C element into the system. That is, power flow can be positive or negative. This does not affect bond graph development; sign conventions will be introduced later.

To be classifiable as a C effect, an action or device should follow the concept that the flow into a C effect causes effort build up within it.

5.2.8 Inertive Power Storage

Inertive power storage occurs when inertias are accelerated, storing up kinetic energy. The power bond graph symbol is:

I

with the power bond (the line) indicating the means through which power may be given to the inertia, or given up by it. Adding the effort and the flow variable associated with the possibilities

. linear motion . rotary motion

leads to

$$\begin{array}{c} Im \\ Fm \,\big|\, \dot{X}m \\ \text{(linear motion)} \end{array} \qquad \text{or} \qquad \begin{array}{c} Ij \\ Tj \,\big|\, \omega j \\ \text{(rotary motion)} \end{array}$$

where Fm is the net force applied to mass Im to cause dynamic linear velocity $\dot{X}m$ of the mass;

Tj is the net torque applied to rotating inertia Ij to cause dynamic angular velocity ωj .

The relationship implied on the I power bond is, of course, Newton's law. However, it is best expressed as

$$\dot{X} = I(F) \qquad \text{or} \qquad \omega j = I(Tj)$$

rather than the more usual

$$Fm = I(\dot{X}) \qquad \text{or} \qquad Tj = I(\omega j)$$

In fact, when we realize that force (torque) is cause, and velocity (angular velocity) is effect, it is more appropriate to express Newton's law as:

$$\dot{X}m = I(Fm) \qquad \text{or} \qquad \omega j = I(Tj)$$

and this leads to

$$\dot{X}m = 1/Im \cdot \int Fm \, dt + \dot{X}m(0) \tag{5.9}$$

or

$$\omega j = 1/Ij \cdot \int Tj \, dt + \omega j(0) \tag{5.10}$$

as the dynamic relationships implied on the respective Im and Ij power bonds.

Inertia effects are referred to as "power storage". As with capacitive effects, I effect power flow can be positive or negative, with the inertia absorbing or giving up power respectively. This will be discussed in Section 5.4.

Electrical inductance is another dynamic I effect. It is described by a dynamic equation similar in form and nature to Newton's law, i.e.

$$E = L \cdot di/dt , \quad \text{or} \quad i = 1/L \cdot \int E\, dt$$

where voltage E and current i are clearly the effort and flow variables respectively, and are respectively analogous to force and velocity in Newton's law.

The general concept of an I effect is that flow is dependent on the time integral of effort.

5.2.9 Summing Junctions

A power bond graph, like all signal flow diagrams, requires summing junctions at which variables can be added algebraically. A power bond graph is, however, a dual signal diagram —— each power bond in it "contains" the two simultaneous variables whose product describes the power flow in the bond. As only like variables can be added to or subtracted from each other, it is necessary to have two distinct kinds of summing junctions in power bond graphs. One of these, depicted by 0 and called an O-JUNCTION, recognizes summation of the flow variables, as follows:

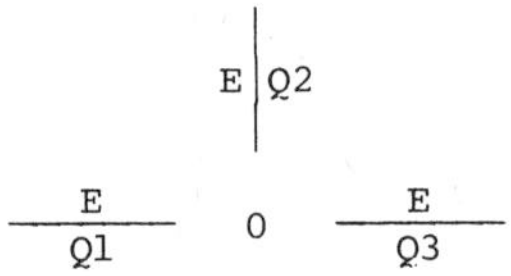

The O-junction means simply

$$Q1 + Q2 + Q3 = 0$$

E = same in all three bonds.

A Tee-junction in a hydraulic line is a typical situation requiring use of an O-junction in a power bond graph,

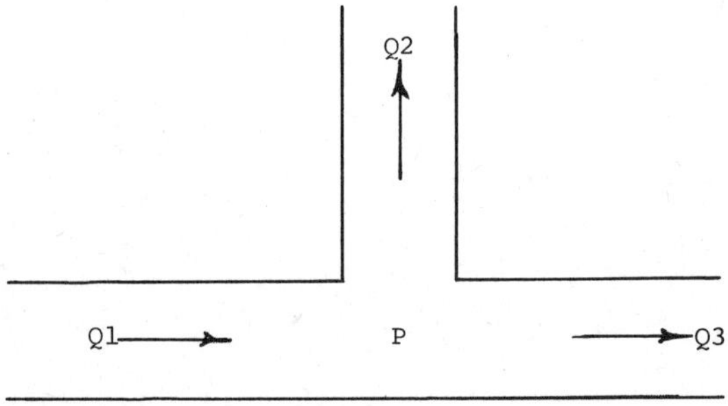

It is obvious that at the Tee-junction, pressure is common. Thus we have

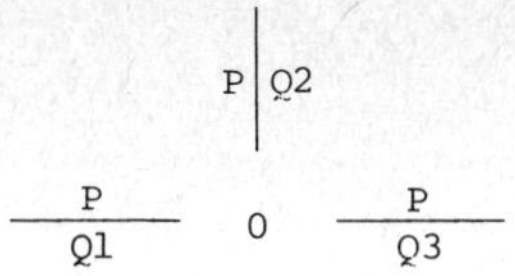

which means

Q1 + Q2 + Q3 = 0

P = same in all three bonds

to represent what is happening at the Tee.

The other summing junction, represented by 1 and called a 1-JUNCTION, recognizes algebraic summation of effort variables at common flow, and is typically

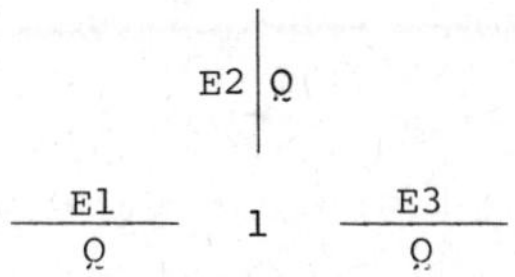

which means

E1 + E2 + E3 = 0

Q = same in all three bonds.

A typical use of the 1-junction occurs when we wish to allow for a frictional resisting force opposing the motion of a hydraulic cylinder-driven load, as follows:

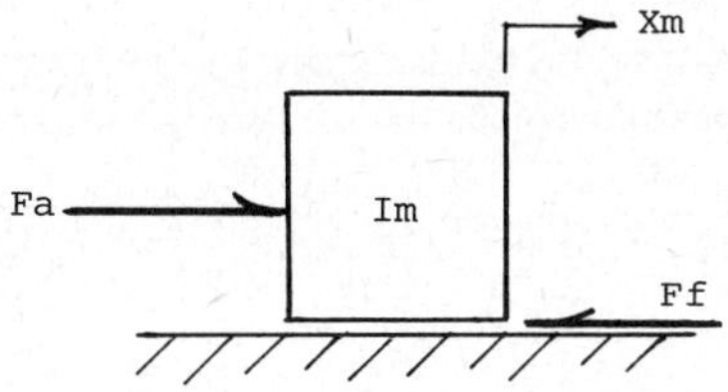

where Fa is pressure induced force acting on mass Im to induce displacement Xm ;
Ff is friction force.

The power bond graph representation,

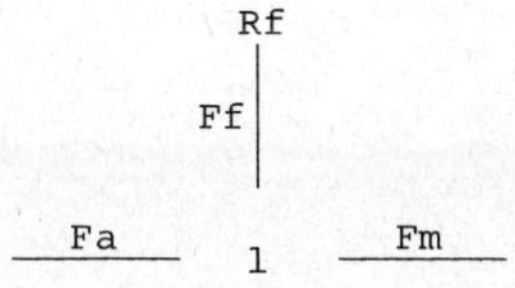

is intended to express

$$Fa - Ff = Fm ,$$

the net force acting on the mass. It is apparent enough that the piston rod, the mass, and the friction force all experience the same velocity $\dot{X}m$, leading to the completed 1-junction,

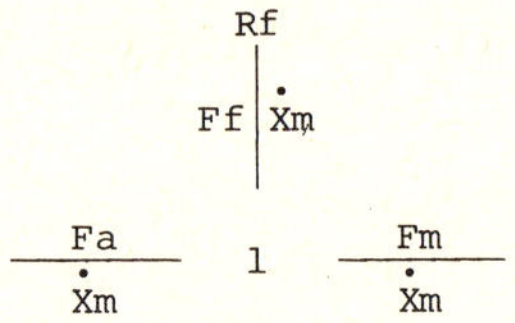

which means, algebraically,

$$Fa + Ff + Fm = 0$$

$$\dot{X}m = \text{same for all bonds}$$

$$Ff = Rf(\dot{X}m) \quad \text{or vice versa.}$$

The method of determining the signs of the terms being added at summing junctions will be illustrated shortly.

5.2.10 Summary of Basic Terms and Symbols

Sections 5.2.1 to 5.2.8 cover the basic terms and symbols needed to form a structure of a power bond graph model of a hydraulic control system.

Fig. 5.1 shows the terms and symbols in summary form.

5.3 FORMING POWER BOND GRAPH STRUCTURES

Before proceeding, let us consolidate use of the basic bond graph terms and symbols by developing bond graph structures for several well known hardware components.

5.3.1 Inertia Load with Friction

It is common that the load driven by a hydraulic control system is in this category. Fig. 5.2 shows the development of a bond graph for an inertia load with friction. Some points to note are:

Step (a): the load system has to receive power as a force • velocity product.

Step (b): the absorption by friction of some of the actuating power is recognized

TERM	SYMBOL	IN BOND GRAPH	MEANING AND COMMENT
Power Bond	——		Power flow in the bond equals the product of E and Q .
Effort	E	E over bond, Q under bond	Effort means potential.
Flow	Q		Flow means flowrate.
Source	SE	SE over bond, Q under bond	Effort is constant, or independent.
	SQ	E over bond, SQ under bond	Flow is constant, or independent.
Transformer	TF	E1/Q1 —— TF —— E2/Q2	E1 transforms to E2 , Q1 transforms to Q2 , such that E1·Q1 = E2·Q2 .
	GY	E1/Q1 —— GY —— E2/Q2	E1 transforms to Q2 , Q1 transforms to E2 , such that E1·Q1 = E2·Q2 .
Resistive Effect	R	—— R	Recognizes a resistive effect.
Capacitive Effect	C	—— C	Recognizes a capacitive effect.
Inertive Effect	I	—— I	Recognizes an inertive or inductive effect.
0-Junction	0	E/Q1 —— 0 —— E/Q3; vertical bond E | Q2	Algebraic summing of flow variables at constant effort, i.e. Q1 + Q2 + Q3 = 0 , E = constant .
1-Junction	1	E1/Q —— 1 —— E3/Q; vertical bond E2 | Q	Algebraic summing of effort variables at constant flow, i.e. E1 + E2 + E3 = 0 , Q = constant .

FIG. 5.1 SUMMARY OF BASIC BOND GRAPH TERMS, SYMBOLS, OPERATIONS

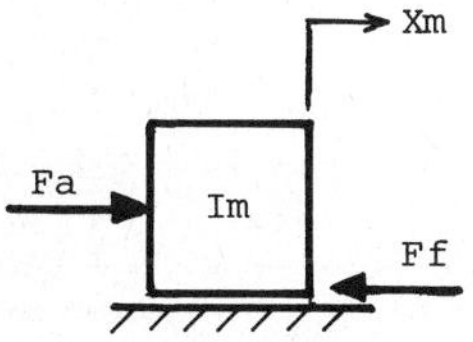

	ACTION
	(a) Power applied to load as force · velocity.
	(b) Friction force recognized by Rf at 1-junction. Fm left as net force.
	(c) Fm acts on load inertia Im (mass). Structure completed.
	(d) Power product symbols added to each bond.

BOND GRAPH DEVELOPMENT

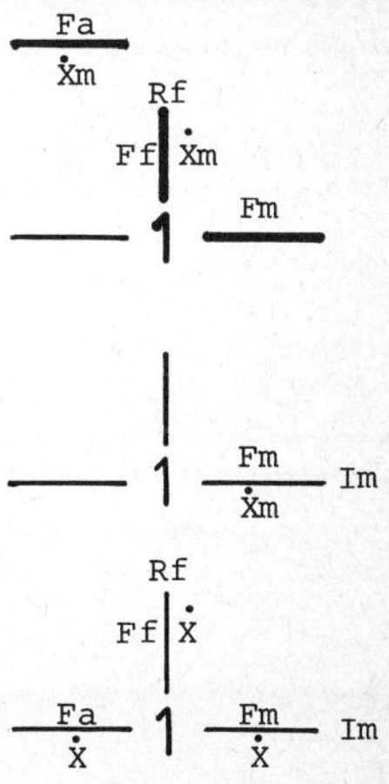

FIG. 5.2 DEVELOPMENT OF BOND GRAPH FOR INERTIA LOAD WITH FRICTION

by the resistive effect Rf , at a 1-junction to show that it is a force summation. Net force Fm emerges.

Step (c): Fm acts to accelerate the load inertia; this is recognized by placing Im at the end of the Fm bond.

Step (d): the two power symbols appropriate to each bond are included on the structure.

Formation of equations from the bond graph will be dealt with later in the Text.

5.3.2 Hydraulic Cylinder

The power cylinder transforms hydraulic power (pressure · flowrate) to mechanical power (force · velocity of ram) . For its dynamic model, it is decided to include friction (piston head and gland seals) and the capacitive effect of the cylinder end volume. For the present, the effect on dynamic response of back pressure on the low-pressure side of the piston will be neglected.

Fig. 5.3 illustrates steps in the development of a bond graph structure. Points to note are:

Step (a): it is recognized that the cylinder will receive power as the product of pressure and flowrate of oil, P · Q .

Step (b): it is recognized that the cylinder will provide power as the product of force Fa and velocity $\dot{X}a$. The implication is that the cylinder *will* be connected to a resisting load.

Step (c): it is considered that the capacitive effect of the cylinder chamber receiving flow is dynamically significant. A certain (small) portion Qc of the inflowing oil is absorbed in compression of the oil in the cylinder. Thus, the effect requires summation at an 0-junction and is designated as a capacitive effect Ca .

Step (d): pressure P transforms to force F on the piston, and flowrate Qa to piston velocity $\dot{X}a$. The area A of the piston is the transformation modulus.

Step (e): the presence of a frictional force Ffa opposing motion of the ram, and due to piston seals and rod glands, is recognized and allowed for as the Rfa bond from a 1-junction.

Step (f): the inertia Ia of the piston and rod will absorb some power as it is accelerated. This is allowed for by the Ia bond; it leads from a

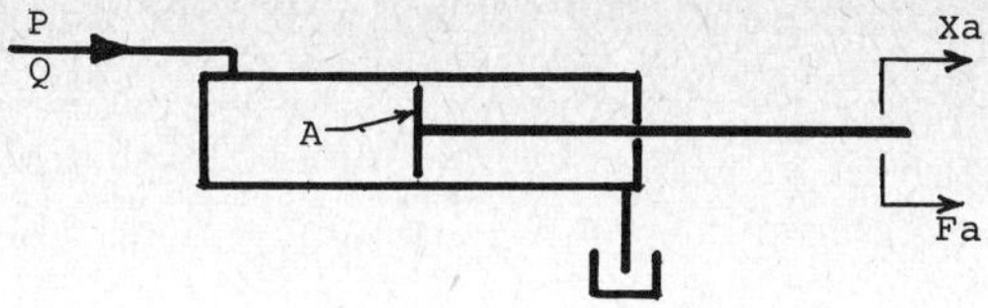

ACTION	BOND GRAPH DEVELOPMENT
(a) Power received as pressure · flowrate.	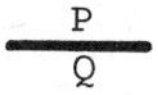
(b) Power output will be as force · velocity.	
(c) The capacitive effect of cylinder volume is recognized, as the Ca bond from an 0-junction.	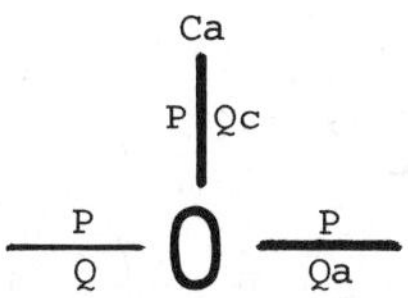
(d) Hydraulic power is transformed to mechanical power. Piston area A is the transformation modulus.	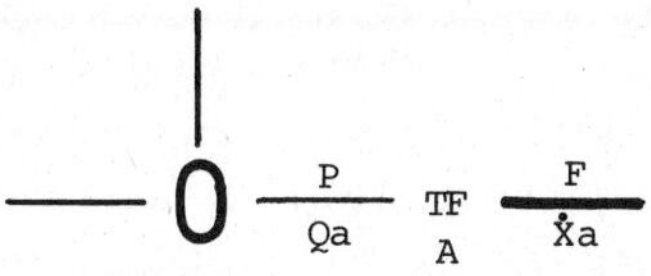
(e) Friction is allowed for by Rfa at a 1-junction.	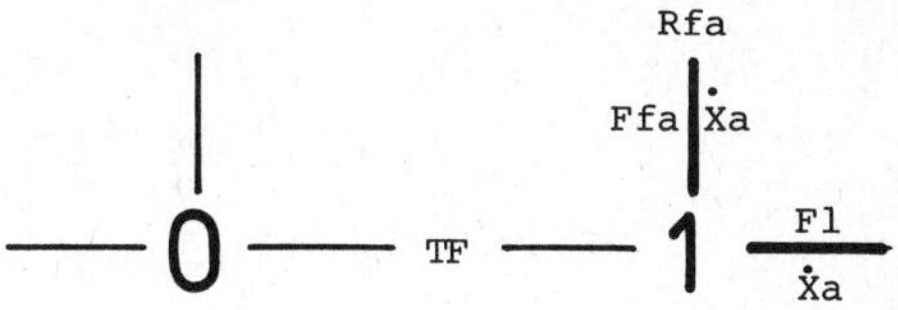
(f) Inertia of ram is allowed for by Ia at a 1-junction. This completes the structure.	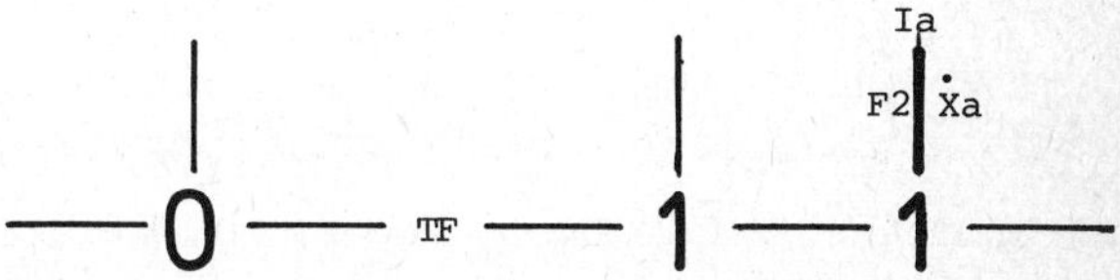
(g) Power variables added to each bond.	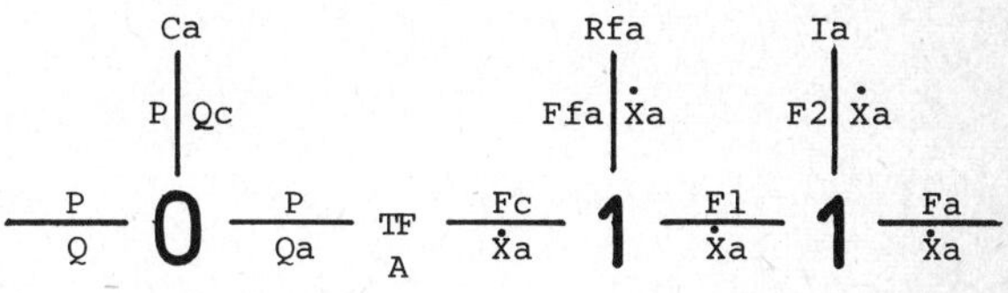
(h) Simplification if cylinder inertia Ia and friction Rfa effects are neglected.	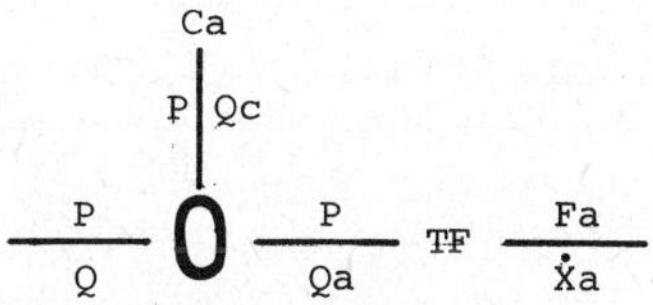

FIG. 5.3 DEVELOPMENT OF A BOND GRAPH STRUCTURE FOR A HYDRAULIC CYLINDER

1-junction as it is force (effort variable) F2 which is being allowed for. The designated effects have all been accounted for. An output bond is joined to the last 1-junction to indicate the actuator will be joined to a load of some kind. This completes the structure.

Step (g): Appropriate power product variables are added to each bond.

The bond graph structure developed for the cylinder is reasonably detailed. It can be simplified by neglecting the cylinder rod inertia and friction (they can often be lumped with the inertia and friction of the load attached to the rod). Such simplification means neglect of the Rfa and Ia effects on Fig. 5.3g in which case the bond graph structure becomes Fig. 5.3h.

5.3.3 Induction Electric Motor

Before a dynamic model is formed, or as it is being formed, decisions have to be made as to what effects are to be included in the model. For the present example, we decide that motor (speed) slip, inertia of the rotor and shaft assembly, friction in the bearings and windage in the rotor, and compliance of the motor shaft all affect the dynamic performance of the induction motor.

Fig. 5.4 shows the steps in development of a bond graph structure, starting with a constant voltage (SE) electric power supply, and ending with a torque (To) and speed (ωo) output power product. Points to note are:

Step (a): supply voltage is recognized as constant (Source SE).

Step (b): it is recognized that the motor gives out power as the mechanical product torque (To) · speed (ωo) .

Step (c): the gyration (GY) of constant voltage to constant motor field speed (Source Sω) is recognized.

Step (d): the speed slip essential to the action of an induction motor is recognized by the Rω bond from an 0-junction. It is a resistive effect.

Step (e): torque loss due to friction is allowed for by the resistive Rf bond leading from a 1-junction.

Step (f): absorption of some torque to accelerate the motor rotor and shaft assembly is allowed for by the Ix bond leading from a 1-junction.

Step (g): the torsional compliance of the output shaft is allowed for with the Ck bond leading from an 0-junction, recognizing that some speed

ACTION	BOND GRAPH DEVELOPMENT

(a) The motor draws power from constant voltage supply.

(b) It gives out power as torque and speed at its shaft end.

(c) Electric power transformed to mechanical power. It is gyration, as SE induces Sω , the synchronous field speed. Ti is ideal torque.

(d) Speed loss due to slip. Rω describes slip characteristic.

(e) Torque lost to friction, in bearings, etc. Rf recognizes friction.

(f) Torque absorbed accelerating motor-shaft inertia Ix .

(g) Speed variation due to compliance of motor-shaft, Ck . This completes structure.

(h) Structure with power state variable symbols added.

(i) Simplified structure, if shaft compliance Ck , rotor inertia Ix , rotor friction Rf are all neglected, and motor field speed ω is regarded as constant (i.e. as a source Sω).

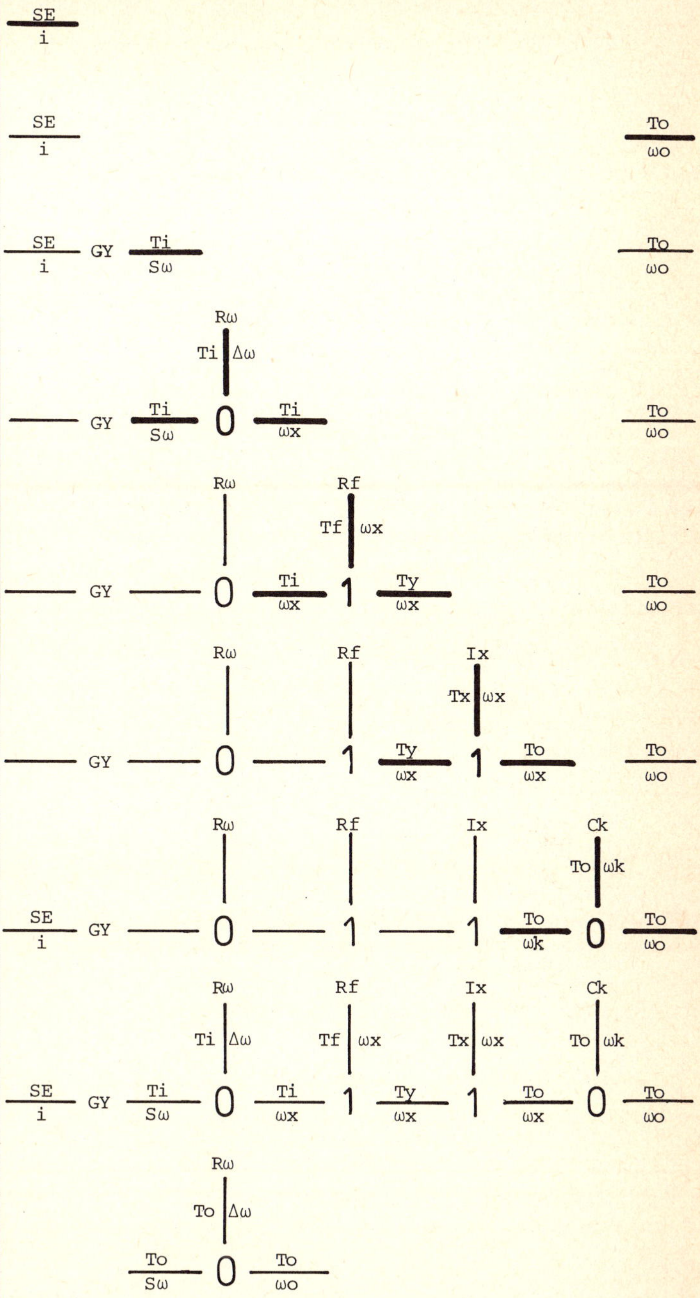

FIG. 5.4 DEVELOPMENT OF BOND GRAPH FOR INDUCTION ELECTRIC MOTOR RECOGNIZING SLIP (Rω), FRICTION (Rf), INERTIA (Ia), AND SHAFT COMPLIANCE (Ck).

fluctuation along the shaft is probable in dynamic operation. The bond structure is now complete, the output bond To · ωo from Step (b) fitting to the Ck 0-junction. The implication is that the To · ωo bond will be joined to the component or system which the motor will drive, for example a hydraulic pump.

Step (h): dual variables appropriate to each bond have been added. These are the variables which will appear in the dynamic equations.

It can be seen that the bond graph comprises a main power flow sequence from left to right consisting of power bonds, transformer, 0-junctions and 1-junctions. From the main stream, power bonds lead from the junctions to the various R , C and I effects judged to affect the motor dynamics.

As far as its effects on a hydraulic system which it might be driving are concerned, the bond graph of Fig. 5.4 is quite detailed. It is likely that for many applications, some of the included effects are insignificant. For example, consider the situation when a hydraulic system is being driven by an electric motor of adequate power, and the analyst's main interest is to predict the dynamic response of the driven inertia load. He will probably decide that the electric motor's inertia, its friction, and its shaft compliance all have negligible effects on load motion. On Fig. 5.4, this means eliminating the Rf , Ix , and Ck bonds and their functions (which are meaningless without the bonds). In addition, the analyst is unlikely to be interested in describing the gyration between voltage and motor field speed, and would start his structure from the constant field speed as a source. In these circumstances, the detailed bond graph of Fig. 5.4h can be reduced to Fig. 5.4i. The only dynamic effect contributed by the electric motor under these circumstances is its speed slip characteristic.

5.3.4 A Simple System

Consider the hydraulic cylinder driven load combination shown on Fig. 5.5a. The bond graph structures for each of the two major components, previously developed, are shown in Fig. 5.5b and c, directly beneath the appropriate hardware components. Actuator inertia and friction are neglected, knowing that they can be lumped with the load's values in this rigidly connected situation.

The system bond graph structure is formed by simple joining of the component bond graphs as shown in Fig. 5.5d. The Fa · $\dot{X}m$ bond directly represents the cylinder rod which connects cylinder piston to load. The ease with which the component structures are joined to form the system structure is a characteristic of the power bond graph technique.

a. The System

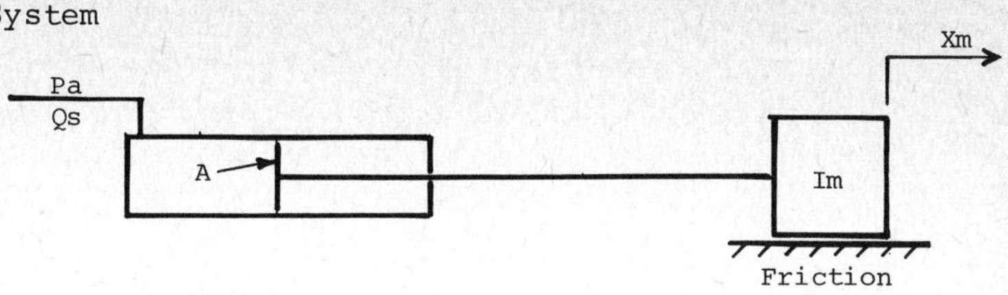

b. Cylinder Bond Graph (Inertia, Friction, and Return Line Effects Neglected)

c. Load Bond Graph

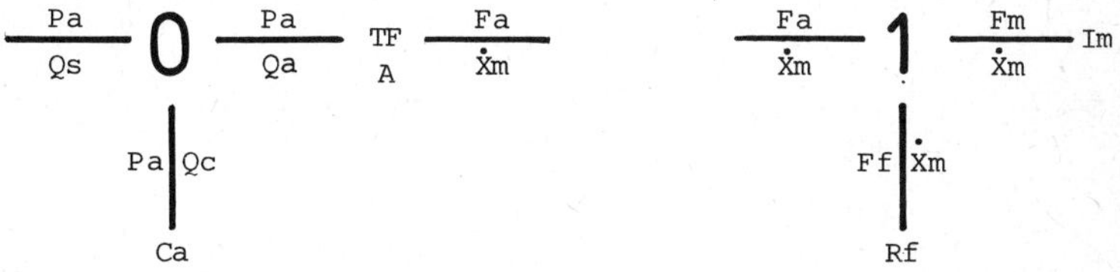

d. System Bond Graph, Formed by Joining b and c at their Common Bond, Fa · Ẋm

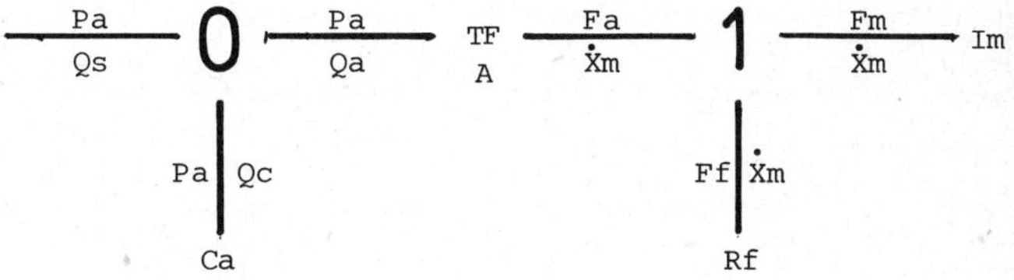

e. System Bond Graph with Power Flow Direction and Causality Assigned (Ref. Section 5.4.6)

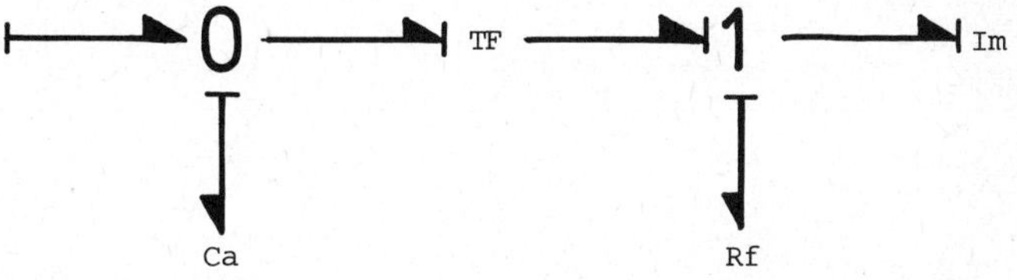

FIG. 5.5 BOND GRAPH FOR SIMPLE CYLINDER-LOAD SYSTEM

More detailed bond graph structures of the linear actuator driven load will be developed in Chapters 7 and 8. Later structures will allow for return line effects, compliance in the actuator-load connection, internal cylinder leakage, and close-centre, open-centre, and tandem-centre valve control.

Fig. 5.5e shows the bond graph with power flow directions (half arrow on each bond) and causality (short transverse bar across one end of each bond) decided. The meanings of these decisions will be presented next.

5.4 POWER FLOW DIRECTIONS, AND CAUSALITY

5.4.1 Introduction

So far in Chapter 5, we have described the basic power bond graph terms and symbols, and have used them to form structures for some components or sub-systems of the kinds included in typical hydraulic control systems. Later, we shall develop more sub-system structures, and also structures for complete systems. A power bond graph structure recognizes the presence of the various R , C and I effects which dictate the dynamic response of the system. The structure does not indicate the relationships by which the dynamic effects are described. That comes later. However, the structure is useful only if it is leading towards eventual preparation of a set of dynamic equations which describe the system dynamics. The assigning of power flow directions, and of causality, are steps preparing the bond graph to yield the appropriate set of equations. Assignment of power flow directions provides the signs of the terms in the algebraic summing equations to be eventually written for each 0-junction and 1-junction. Assignment of causality reflects decisions on which power state variable is to be cause, and which is to be effect, on each R , C and I bond and its equation.

5.4.2 Directions of Power Flow

As with all dynamic modelling procedures, a sign convention must be adopted so that the signs of terms in equations are compatible with each other and with system action. As with other flow diagrams, a conventional flow direction is indicated by arrowheads. This does not imply that flow is always in the designated direction. Some flows may be positive or negative at different times during the dynamic action. However, if the signs of the equation set are correct for one (any) instant, they will be correct globally, i.e. for the entire action. Adoption of a sign convention, and sticking to it, acts to ensure that the signs of terms in the model equations are correct. Common sense should be utilized in the designation of a sign convention — the convention should reflect real system behaviour as far as is possible.

A realistic approach to getting signs correct is:

- imagine the system to be in equilibrium, at time $t = 0$,
- imagine the system to be activated at $t = 0$,
- decide signs by the physical action which takes place in the system immediately after it is activated.

With bond graphs, it is power which is flowing in each bond. It becomes necessary to indicate the direction of power flow in each bond. The convention adopted is that power is shown as entering each R , C and I element by a half arrowhead as follows:

The directions of power flow in the remaining bonds of a bond graph follows physical common sense —— power flows from the source to the load, usually from left to right across the bond graph.

Consider the bond graph structure developed for the hydraulic cylinder in Fig. 5.3, and shown again on Fig. 5.6. Obviously, it is intended that power flows from left to right, i.e. from the hydraulic supply $P \cdot Q$ towards the driven load. Hence, directional half arrows are added on the main (horizontal) power flow path from left to right. For Ca , Rf and Ia , show the half arrows pointing into the effect, utilizing the convention that power flows into these elements.

It is usually a simple matter to designate the power flow direction on each bond. For the present, this merely helps in appreciating what is going on in the system. Later we shall use power flow directions in deciding the signs in the equations being formed from the structure.

Technically, designation of power flow in a certain direction reflects a decision that the flow variable is directed in that direction at the instant considered. But keep in mind that we are concerned with dynamic effects, i.e. changes from equilibrium states rather than steady-state values of variables. A flowrate can drop from one positive value to another lower positive value; the dynamic change is negative, but the physical flow remains positive!

5.4.3 Causality

As discussed in Chapter 4, consideration of whether a power state variable is a cause variable or is an effect variable gives insight into the dynamic nature and behaviour of the system. The idea of *causality* —— i.e. the designation of a power

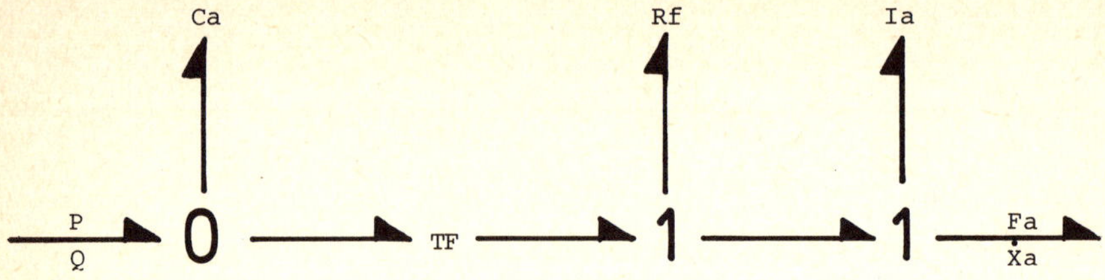

FIG. 5.6 BOND GRAPH OF HYDRAULIC CYLINDER (EX. FIG. 5.3), WITH POWER FLOW DIRECTIONS ADDED

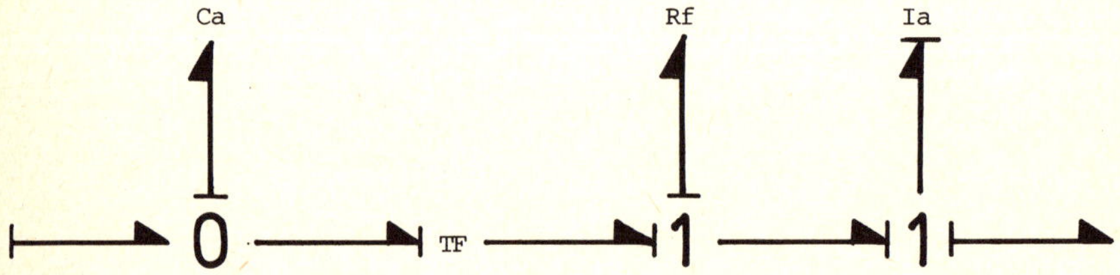

FIG. 5.7 BOND GRAPH OF HYDRAULIC CYLINDER (EX. FIG. 5.6), WITH CAUSAL BARS ADDED.

variable as a cause —— can be extended and applied to a power bond graph structure so as to show how the equations to be prepared for R , C and I effects are to be arranged. Basically, it is most appropriate to write effect = f(cause) ; for subsequent digital simulation, it is preferred that a differential relationship be expressed in integral rather than derivative form.

A causal decision is indicated by a short transverse bar, called the *causal bar*, drawn across one end of a power bond. For example,

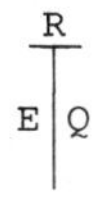

with the causal bar drawn at the R end of the power bond, indicates that the R expression is to be written in the form

$$Q = R(E)$$

where R() indicates "a resistive function of (what is in parenthesis)". That is, it has been decided that E is the cause and Q is the effect for this particular R element.

If the causal bar is placed on the power bond at the end away from R ,

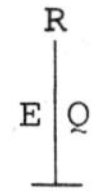

this is a statement that the R equation is to be prepared in the form

$$E = R(Q)$$

which says that Q is the cause, and E the resulting effect.

In short, a causal decision on an R , C or I power bond indicates which of the dual power variables on the bond is to be used to compute the other. The object of the equation —— i.e. the left-hand side in this Text —— is the effect variable, the subject (the variable in the right-hand or functional side of the equation) is the cause.

Causal assignment should reflect reality both as regards to how the system components behave and how particular equations should be arranged. For example, computer solution (simulation) procedures prefer equations in integration forms rather than in differential forms. Thus, capacitive equations should be expressed:

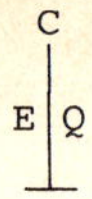

with the causal bar away from the C end of the power bond, leading to equations such as

$$E = C(Q) = \text{(possibly) } 1/C \cdot \int Q\, dt + Eo$$

instead of the differential form

$$Q = C(E) = \text{(possibly) } C \cdot \dot{E}$$

which would be implied if the causal bar was at the C end. Note that C has been used in the above expressions to indicate a capacitive effect, a capacitive function of, and in the possible equations as a constant.

On the other hand, inertive equations should be expressed:

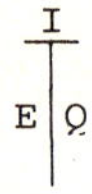

with the causal bar at the I end, leading to Newton's law being expressed as

$$Q = I(E) = 1/I \cdot \int E\, dt + E(0) \qquad (\text{or } \dot{X}m = 1/Im \cdot \int Fnet\, dt + \dot{X}m(0)\ )$$

rather than in the differential form

$$E = I(Q)\ I \cdot \dot{Q} \qquad (\text{or } Fnet = Im \cdot \ddot{X}m\ )$$

which is implied if the causal bar is away from the I .

For the reasons given, C and I elements on a power bond graph are normally given preferred causalities, the causal bars being away from a C element, and adjacent to an I element. R elements can have either causality, depending upon the particular R and on causalities flowing from the preferred C and I decisions. The total causal scheme for a system must make sense. One decision will affect others.

Causality can be interpreted that the effort variable is directed *at the causal bar*, and consequently, that the flow variable is directed *away* from the causal bar. For example,

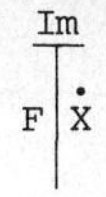

means that force F is directed at the mass Im and velocity $\dot{X}$ is a resulting (back) effect (which is not to say that velocity is backwards, or negative!). Conversely, for a hydraulic accumulator,

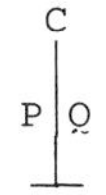

means that flowrate Q is directed into the accumulator C and pressure P is a resulting (back) effect. That is, pressure P is directed at the causal bar, away from C . This physically sensible interpretation helps in completion of the causal decisions for a system, as it dictates the causal bar state at each of the summing junctions (0 and 1-junctions).

For example, there can be only one effort variable at an 0-junction, as previously demonstrated, indicating that there can be only one causal bar at an 0-junction,

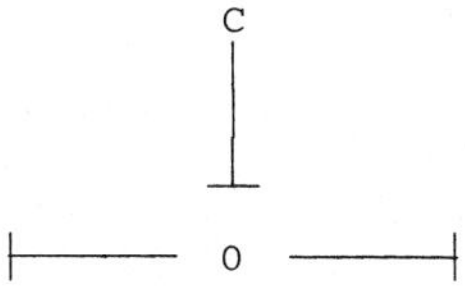

If, as in the above case, a C element occurs at an 0-junction, the other bonds at the junction *must* have causal bars away from the junction.

On the other hand, there can be only one flow variable at a 1-junction, as previously demonstrated, and this means that there can be only one bond at a 1-junction without a causal bar at the junction. Representation of an I element demonstrates this

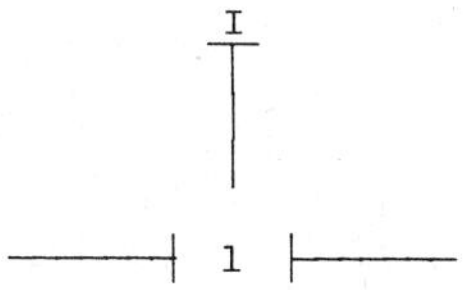

With the preferred causal bar against the I end of the I bond, the remaining two power bonds *must* have causal bars at the 1-junction if the symbolic representation is to make physical sense.

Finally, assigning causality is a quite separate operation from assigning power

flow directions.

5.4.4 Example: Hydraulic Cylinder

As a more complete example, consider the bond graph structure for the hydraulic cylinder from Section 5.3.2 and Figs. 5.3 and 5.6. Fig. 5.7 shows the bond graph with causality added, the required steps being:

1. Preferred Causalities: place causal bar away from Ca , and against Ia .

2. Consequent Causalities: at the Ca 0-junction, the two horizontal bonds *must* now have causal bars away from the junction. Add them.

   At the Ia 1-junction, both horizontal bonds *must* now have causal bars at their junction ends.

3. Remaining Causalities: at the Rf 1-junction, both the Rf and TF bonds *must* now have causal bars at the 1-junction.

Thus, simply using the C and I preferred causalities, the remaining causalities fell into place. It is not always so unambiguous, but more of that later.

5.4.5 Example: Induction Electric Motor

Fig. 5.4 showed the development of a bond graph structure for an induction electric motor. Fig. 5.8a shows it with power flow directions assigned. Fig. 5.8b shows the bond graph with causality assigned. Considering Fig. 5.8a,

1. Assign preferred causalities to Ck and Ix , by placing a causal bar at the 0-junction end of the Ck bond, and at the Ix end of the Ix bond.

2. The causal bars of the other two bonds at the Ck 0-junction must now be placed at the bond ends away from the 0-junction.

3. The causal bar of the left bond at the Ix 1-junction now must be at the 1-junction.

4. Because the right-hand bond at the Rf 1-junction cannot now have a causal bar at the junction, the remaining two bonds must have causal bars at the junction.

   This leaves the gyrator bonds and the Rω 0-junction as causally undecided, except that the right-hand bond from the 0-junction cannot have a causal bar at its junction end. Rational thought lets us resolve this.

a. Power Flow Directions Decided

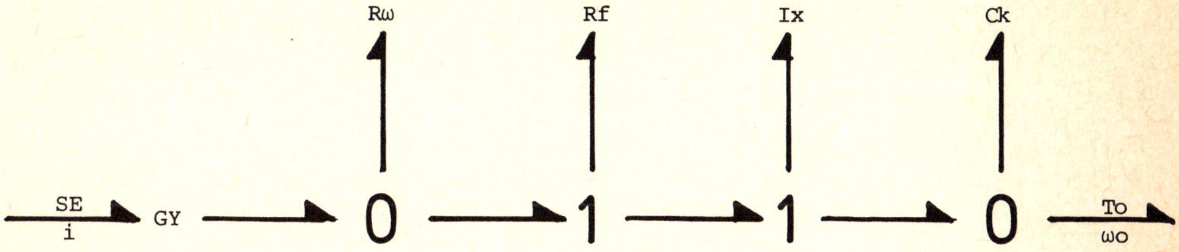

b. Causality Decided

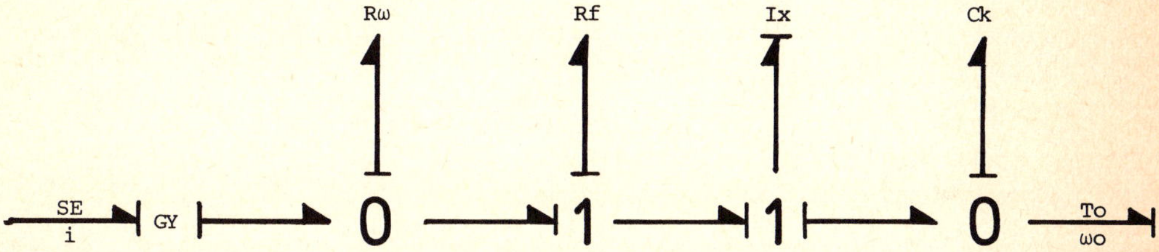

FIG. 5.8 BOND GRAPH STRUCTURE FOR INDUCTION ELECTRIC MOTOR (EX. FIG. 5.4)

5. A fundamental characteristic of an induction motor is that load causes speed slip, and slip induces torque to drive the load. That is, as far as the torque-slip characteristic is concerned, slip is the cause and torque is the resulting effect. Therefore we should express their relationship as

$$T = R(\omega)$$

by placing the $R\omega$ bond's causal bar at its 0-junction end. This immediately locates the causal bar for the adjacent left bond at its GY end, leaving only the SE/i bond causality undecided.

6. Voltage SE is a constant, unaffected by the system. It is also the effort variable. Effort variables are directed at the causal bar. Thus SE must be directed into the system and its causal bar should be placed at the GY end of its bond.

The causal assignment is now complete, and the bond graph is ready to have the dynamic equation set prepared from it.

5.4.6 Example: Simple Cylinder-Load System

Consider the system of previous Fig. 5.5 whose bond graph structure was shown on Fig. 5.5d. Using only the criteria

- power flows into R , C , I elements,
- power flows from source to load,
- causal bar at the I end of I bonds,
- causal bar at the 0-junction end of the C bonds,
- only one causal bar at an 0-junction,
- causal bars on all bonds except one at 1-junctions,

the power flow directions and causal bars of Fig. 5.5e are simple to complete.

5.4.7 Summary

The reasons and the techniques for assigning power flow directions and causality have been presented and discussed. Their significance will be confirmed in Section 5.5, where equations will be developed from power bond graph structures. However, patterns which reflect both computational requirements and physical reality have emerged, and will be summarised here.

- Power flow direction and causality decisions are intended to help in the processing of a rational set of dynamic equations from a power bond graph structure.

- Given a bond graph structure, allot each C and I element its preferred causality; C effects have the causal bar placed away from the C ; I effects have the causal bar placed against the I .

- Assign appropriate causality to any designated sources.

- Use the physical facts that

  - an 0-junction can have only one causal bar adjacent to it,

  - a 1-junction can have only one bond without a causal bar adjacent to the junction,

  to decide causalities consequent upon the C , I , and S causality decisions.

- If all remaining causalities do not flow automatically from the above steps, seek the most physically obvious causal decision(s) to allow completion of causal assignments.

- Each power bond must have a causal bar, at one end only.

- If causal conflicts arise, they must be resolved. If they cannot be resolved, there is something wrong with, or missing from, the power bond graph structure. *This is a key check against validity of the proposed model.* It does not necessarily follow that a bond graph without causal conflict is complete. As with all dynamic modelling procedures, experience and engineering judgement enhance the probability of success.

- It is best to decide causalities on the complete system bond graph, rather than on each component bond graph before they are put together. Causality of a component is not necessarily unique, and may have to be re-thought when the component goes into a specific system. Indeed, the bond graph structure of a component may have to be re-thought when it is to be used in a particular system.

- Reflection and experience will confirm that at a TF power transformer there can be only one causal bar

  i.e. ——| TF ——| or |—— TF |——

  but not ——| TF |—— nor |—— TF ——|

  For GY power transformation the opposite holds, i.e. there must be either

TERM	SYMBOL		MEANING AND COMMENT
Power Bond	——		Power flow in the bond equals the product of E and Q .
Effort	E	E —— Q	Effort means potential.
Flow	Q		Flow means flowrate.
Source	SE	SE —— Q	Effort is constant, or independent of the system.
	SQ	E —— SQ	Flow is constant, or independent of the system.
Transformer	TF	El —— TF —— E2 Ql —— —— Q2	El transforms to E2 , and Ql transforms to Q2 such that El·Ql = E2·Q2 .
	GY	El —— GY —— E2 Ql —— —— Q2	El transforms to Q2 , and Ql transforms to E2 such that El·Ql = E2·Q2 .
Resistive Effect	R	—— R	Recognizes a resistive effect.
Capacitive Effect	C	—— C	Recognizes a capacitive effect.
Inertive Effect	I	—— I	Recognizes an inertive or inductive effect.
0-Junction	0	E —— 0 —— E Ql —— —— Q3 E ┃ Q2	Algebraic summing of flow variables at same E , i.e. Ql + Q2 + Q3 = 0 .
l-Junction	1	El —— 1 —— E3 Q —— —— Q E2 ┃ Q	Algebraic summing of effort variables at same Q , i.e. El + E2 + E3 = 0 .
Power Flow Direction	⇀ (half arrow on one end of each bond)	——⇀	Shows direction of power flow in each bond. Easily decided for main power flow sequence. Power flows into R , C , I elements (a convention). Gives sign convention for equations to be formed for junctions.
Causality	┃ (short bar across one end of each power bond)	——⇀┤ or ├——⇀	Shows which of the two variables on a bond is cause and which is effect. Decides how R , C , I equations are to be arranged. Only one causal bar can be at an 0-junction. Only one bond can be at a l-junction without a causal bar. C elements have causal bar away from the C end. I bonds have causal bar at the I end. Effort is directed at the causal bar, flow away from it.

FIG. 5.9 SUMMARY OF COMMON BOND GRAPH TERMS AND SYMBOLS

two or zero causal bars at the GY

i.e. ——| GY |—— or |—— GY ——|

but not ——| GY ——| nor |—— GY |——

Fig. 5.9 shows tabulated the common bond graph terms and symbols. Fig. 5.9 is similar to Fig. 5.1, but with power flow direction and causality included.

5.5 PREPARING EQUATION SET

5.5.1 Introduction

The major objective of forming a power bond graph and assigning power flow directions and causality to it is to provide steps towards the formation of a set of equations whose solution provides the predicted response of the system.

Steps for forming the equations from a bond graph will now be demonstrated, and then summarised into a reasonably formal procedure.

5.5.2 Example: A Simple Component

Consider the bond graph for the inertia load with friction shown on Fig. 5.10 (Fig. 5.2, with power flow and causality assigned). Starting with the dominating inertia effect, the relationships implied on the bond graph are:

1. $\dot{Xm} = I(Fm)$ --- arrangement dictated by causal bar at Im .

   $= 1/Im \cdot \int Fm\,dt + \dot{Xm}(0)$ --- Newton's law provides the actual equation.

2. $Ff = R(\dot{Xm})$ --- arrangement dictated by causal bar being away from Im .

   $= Kf \cdot \dot{Xm}$ --- if viscous friction assumed; Kf is coefficient.

   $= \mathrm{sgn}\,\dot{Xm} \cdot Ffu + Kf \cdot \dot{Xm}$ --- if coulomb friction Ffu is included.

3. $Fa = ?$ --- to be specified if Fa is an external input, or to be described from bond graph structure of the component driving the load, as discussed in the next example.

4. $Fm = Fa - Ff$ --- summing junction equation, Fm being net force on load. Signs satisfy power flow half arrows, i.e. $Fa - Ff - Fm = 0$; Fm chosen as object (left-hand side) of equation as Fa and Ff are already objects of equations.

Equations 1 to 4 are the model, formally derived from the bond graph. Note that

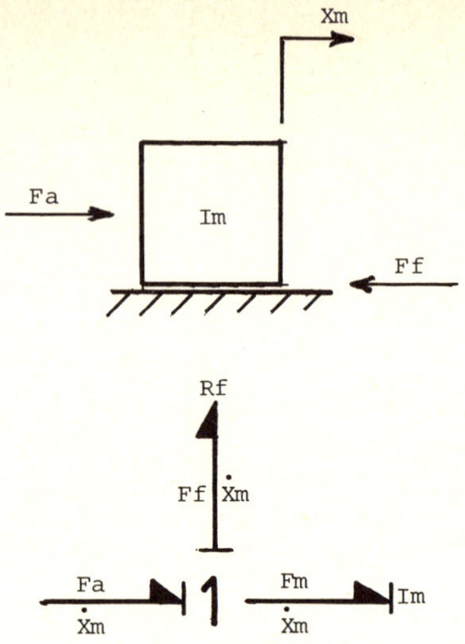

FIG. 5.10 BOND GRAPH FOR INERTIA LOAD WITH FRICTION

a. System

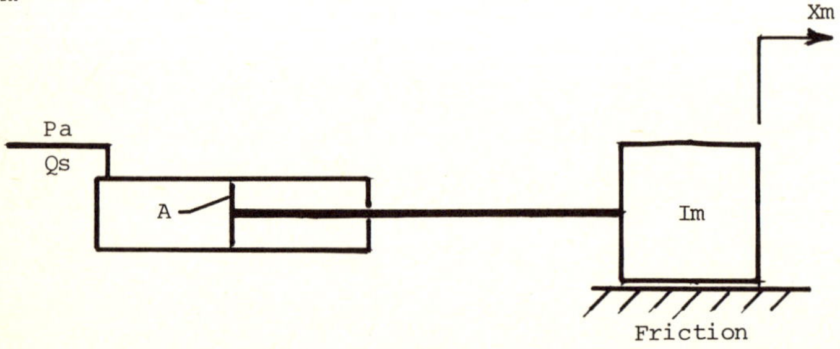

b. Power Bond Graph

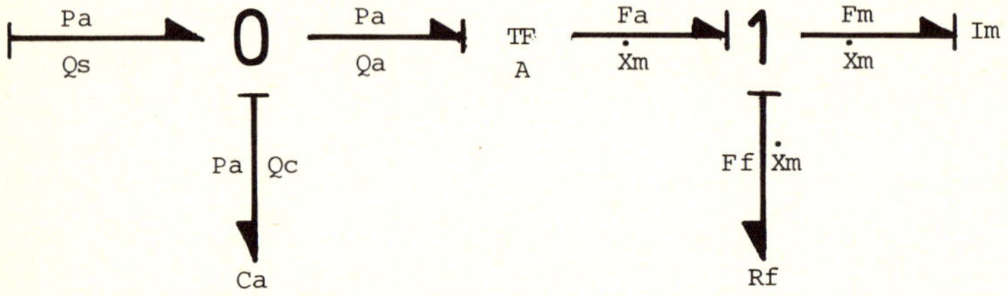

FIG. 5.11 BOND GRAPH FOR CYLINDER DRIVEN LOAD SYSTEM

there are four power state variables associated with the bond graph (Fa , $\dot{Xm}$, Ff , Fm) and that each is the object of a power state equation. It is probable that load displacement $Xm(t)$ would be of primary interest to the analyst; it is therefore added as an auxiliary equation by describing:

5. $Xm = \int \dot{Xm}\, dt + Xm(0)$ --- simple integration of $\dot{Xm}$.

5.5.3 An Elementary System Example

Fig. 5.11 shows the bond graph for a cylinder-driven inertia load with friction. It was derived, without power flow or causality, in Section 5.4.6 and Fig. 5.5. The supply flow of oil to the cylinder must come from an upstream component such as a 4-way control valve. However, for the moment let us keep it simple. As with the previous example, we shall work back from the driven load in preparing the equations:

1. $\dot{Xm} = 1/Im \cdot \int Fm\, dt + \dot{Xm}(0)$ --- Im equation, as previously.

2. $Ff = Kf \cdot \dot{Xm}$ --- Rf equation assuming viscous friction.

3. $Fm = Fa - Ff$ --- net force acting on mass; equation of the 1-junction.

4. $Fa = Pa \cdot A$ --- transformation of cylinder pressure Pa to force Fa via modulus A .

5. $Qa = A \cdot \dot{Xm}$ --- transformation between swept volume rate of piston and piston velocity via modulus A .

6. $Pa = C(Qc)$ --- dictated by causal bar functional arrangement of the Ca equation.

   $= 1/Ca \cdot \int Qc\, dt + Pa(0)$ --- assuming linear capacitance with Ca a coefficient.

7. $Qs = ?$ --- determined by next upstream component.

8. $Qc = Qs - Qa$ --- equation of the 0-junction, signs taken from power flow arrows; Qc chosen as object because Qs and Qa already are objects of equations.

The equation set 1 to 8 is the power state model for the system; each power state variable is the object of one equation; auxiliary equations can be added if it is desired to study other than power state variables, or if some of the R , C , I quantities are other than simple coefficients and require further description. For example, except for small perturbations about a mean value Xm , the capacitance of the hydraulic cylinder is not a simple coefficient Ca . This is because the volume of the cylinder end, and hence of the oil under pressure Pa , changes significantly with Xm . Remember, a C , R , and even I symbol on a bond graph does not necessarily refer to a coefficient, although very often this is the case. More fundamentally, a C refers to "a capacitive function", R to "a resistive function", and I to "an inertial function". A coefficient is, of course, the simplest functional relator between two quantities.

It is informative and significant that the causal bar had to be placed at the input end of the Pa • Qs bond. It conveys that the component directing flow to the actuator is essentially a flowrate provider, and that pressure is a consequent back effect dictated by the actuator and load system. The bond graph reveals the fallacy of thinking that pressure can be *supplied* to an actuator.

5.5.4 A More Complete System

System and Assumptions

Fig. 5.12a shows the system. It is desired to predict load motion Xm(t) following control commands Xv(t) to the valve spool.

For the present analysis, it will be assumed

- that the pumping system supplies oil to the control valve at constant pressure and that pump supply is adequate. This allows neglect of the dynamic effects of all components upstream of the control valve, reducing the hardware to be considered to that shown on Fig. 5.12b. Further, supply pressure to the valve is designated as a source, SP ,

- that the flow in the return lines (cylinder to valve to tank) has a negligible effect on load motion response Xm following valve actuation,

- that the resistance of the high pressure side line is negligible.

The assumptions allow reduction of the load and cylinder bond graphs to those, used previously, shown on Fig. 5.12c under the respective components of Fig. 5.12b. We have not previously required a bond graph for a 4-way valve.

4-Way Control Valve

The closed-centre 4-way (i.e. 4 port) valve has four control orifices each of which can be regarded as a resistance to flow. It is control of these resistances by the operator which provides control of the system. In the central position for an ideal closed-centre valve, each resistance is infinitely high, and each port flow is zero. As the valve spool is moved, control orifice resistances decrease with increasing Xv , falling to minimum values when the ports are fully open. Control valve flows, including open-centre and tandem-centre arrangements, are discussed in detail in Chapter 7.

When return line effects are neglected, only the supply port and the pressure-side control port need be considered. Further, the ports are physically close, and together with the internal passage connecting them, they can be lumped as a single resistance. It is of course pressure that is "lost" during flow through a restriction. Thus, for dynamic analysis purposes, the left-hand bond graph structure of Fig. 5.12c summarises the contribution of the valve to the present system's dynamic action. What we have is

- constant supply pressure SP and its associated supply flowrate Qs on the input bond. Note that Qs is not constant —— it is both pressure-drop and valve-displacement dependent,
- valve resistance accounted for as the Rv bond, leading from a 1-junction to show that pressure drop ΔPv (effort variable) is being accounted for,
- pressure Pa (= SP - ΔPv) and Qs go forward from the valve to the next connected component.

System Bond Graph

Fig. 5.12d shows the bond graph for the three connected components, including power flow directions and causal bars. Using the conventions of causality, only the causal bar locations on the SP · Qs and the Rv bonds were in doubt. At the left 1-junction, prior satisfaction of C and I causalities leaves the causal bar for Pa · Qs at the 1-junction end, as follows,

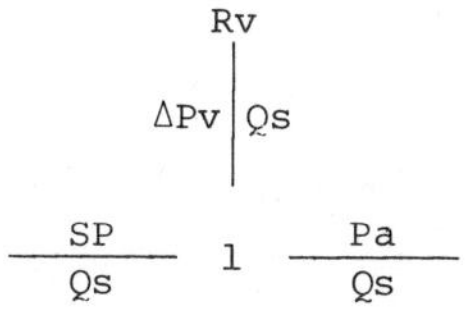

with causality for the other two bonds to be decided. To resolve this, consider the physical reality of either bond. For the Rv bond, for example, it is usual to express flowrate through the valve as a function of pressure drop

$$Qs = f(\Delta Pv, Xv)$$

Thus, the causal bar on the Rv bond needs to be at the bond's Rv end. This means in turn that the causal bar on the SP bond must be at its 1-junction end.

Alternatively, if supply pressure is a source SP , it cannot be a back effect (load dependent). Using the idea that the effort variable is directed at the causal bar, the SP bond causal bar must be in the forward power flow direction, against the 1-junction. This decision, together with the already established causal bar for

a. The System

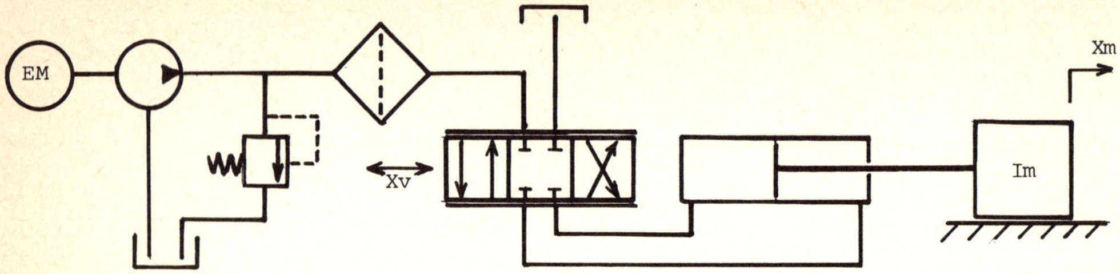

b. The Dynamic Components if Supply Pressure at Control Valve Assumed Constant

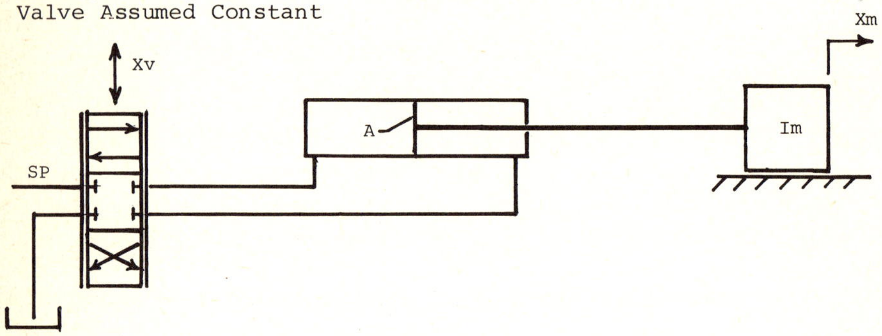

c. Component Bond Graphs, if Return Line Effects Neglected and Xv is either Positive or Negative Only

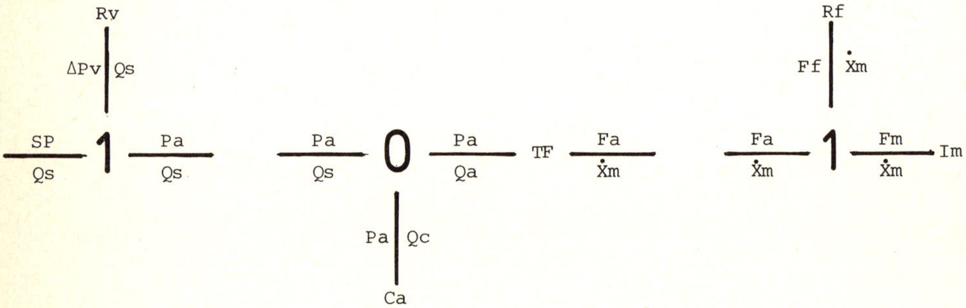

d. System Bond Graph

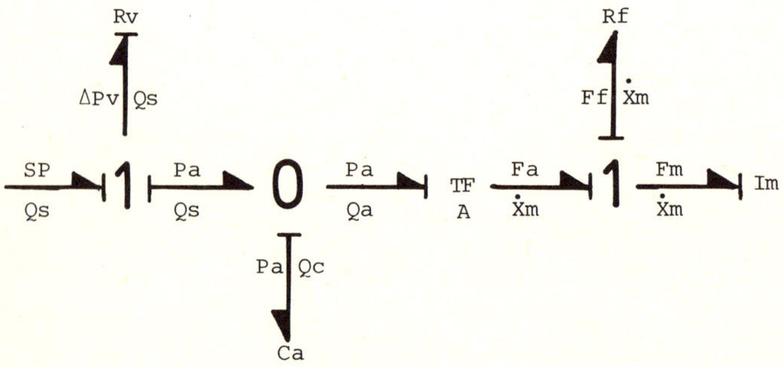

FIG. 5.12 BOND GRAPH FOR A SYSTEM

the Pa · Qs bond, means that the Rv bond causal bar must be at its Rv end.

Note how the need to resolve a doubtful causality gives insight into the system's cause-effect behaviour.

System's Equations

There are ten different power variables on Fig. 5.12d, and so we need ten power state equations. We shall prepare them in the order

- Sources
- I effects
- C effects
- R effects
- TF
- Junctions (0 to 1 from left to right), giving,

Source: $SP = \text{constant}$

Im: $\dot{Xm} = 1/Im \cdot \int Fm\,dt + \dot{Xm}(0)$

Ca: $Pa = 1/Ca \cdot \int Qc\,dt + Pa(0)$

Rf: $Ff = R(\dot{Xm})$. To be specified; e.g.

$= Kf \cdot \dot{Xm}$ for viscous friction, Kf being a coefficient

Rv: $Qs = R(\Delta Pv)$ from causal decision

$= K \cdot Xv \cdot \text{sgn}\,\Delta Pv \cdot \Delta Pv^{\frac{1}{2}}$ (see Chapter 7 for detailed explanation)

TF: $Fa = Pa \cdot A$, $Qa = \dot{Xm} \cdot A$ — arranged to have previously unused object variables

left 1-junction: $\Delta Pv = SP - Pa$

0-junction: $Qc = Qs - Qa$

right 1-junction: $Fm = Fa - Ff$

(arranged to have previously unused object variables, and using power flow direction arrows to decide signs)

as the set of power state equations. Auxiliary equations which will be required are:

$Xv =$ (input to be specified)

$Xm = \int \dot{Xm}\,dt$ (desired primary response variable).

In a specific system, the various independent parameters and coefficients need to be specified numerically (SP , Im , Ca , Kf , K , A , Xv) whereon the equations can be simulated to provide response predictions. Some examples shall be carried through to complete solutions in Chapter 8.

5.5.5 Summary

The examples given can be worked back to include a pressure relief valve, hydraulic pump, an electric motor and other components. However, this will be left to future examples and case studies. The purpose at present is to illustrate how the equation set is formed from a bond graph structure.

There is not an exclusive order of the equations. However, the bond graph novice is advised to seek equations in the following sequence. Having obtained a bond graph with power flow directions and causality assigned,

. write an equation for each source,

. write at least the functional relationship for each R , C , I , with functional arrangement dictated by causality,

. write the effort equation for each TF and GY , at least in functional form, using as the object a variable not yet the object of an equation,

. write the flow equation for each TF and GY , again using as the object a previously unused variable,

. write a summing equation for each 0 and 1-junction, using power flow directions to dictate signs, and choosing as the object of each equation a power variable not already the object of an equation,

. check that there is an equation for each power state variable, and that no such variable is without an equation,

. decide any specific relationships in the power state equations as yet undecided,

. write an equation for each of any system variables required for study which are not among the power state variables,

. write defining equations for any parameters appearing, but as yet undefined, among the power state equations. Typically, a control input might exist (such as control valve displacement Xv).

To get a solution from the model,

. assign numbers to all coefficients appearing in the completed equation set,

. simulate (solve) to get specific system's response.

With some experience gained, an analyst may prefer to prepare the equations from a bond graph in a somewhat different order; typically to start at the first (or last!) summing junction. For the systems of Fig. 5.12d, this would result in

First (left) 1-junction: $SP = \text{constant}$

$$QS = R(\Delta Pv) = K \cdot Xv \cdot \text{sgn}\, Pv \cdot Pv^{\frac{1}{2}}$$

$$\Delta Pv = SP - Pa$$

First (left) 0-junction: $Pa = 1/Ca \cdot \int Qc\, dt + Pa(0)$

$$Qc = Qs - Qa$$

and so on. With this approach, the choice of object for each summing equation is less certain, and may require subsequent re-arrangement of the equations.

The power bond graph technique results in a set of relatively simple equations, readily checked for errors, and readily programmed for computer solution. The same forms of equation appear in example after example, and those relationships relevant to hydraulic control systems quickly become ingrained and are easily recalled and used.

It is possible to prepare from bond graphs equations in the conventional transfer function and state-variable forms. Karnopp and Rosenberg [5.1] describe in detail how to develop the state-variable form $\dot{\bar{X}} = \underline{A} \cdot \bar{X} + \underline{B} \cdot \bar{U}$. This development will not be presented nor encouraged in the present Text as it appears to offer no advantage. Indeed, the structural rigidity required for transfer function and state-variable model forms may be a positive hindrance. Suffice it to keep in mind that models in conventional forms can be formed from bond graphs if there is a compelling reason.

5.6 SOME FURTHER ASPECTS OF BOND GRAPHS

Bond graph methods well beyond what appears in the present Text have been developed and used. The present Text concentrates only on those aspects of bond graph methods which the author has found to be useful and appropriate to the dynamic analysis and performance prediction of industrial hydraulic control systems. Readers wishing to delve further into bond graph techniques can do so via References 5.1 to 5.4. However, there are several aspects of bond graphs not discussed in detail in the present Text which should be brought to the attention of readers.

5.6.1 Modulation of Effects

In some systems, one or more effects may be modulated during system action. The modulation can be independent of the system, or can be made via automatic feedback. In either case, it is usual that the modulation action does not utilize significant system power. It is usual that *one* system variable, rather than a power pair, is utilized to effect any internal system feedback. For example if a variable displacement pump was adjusted manually during operation, no system power is utilized. If a solenoid-operated hydraulic valve is actuated by a system pressure, no power is drawn from the hydraulic system itself. In general, logic, control, and information signals are sensed, processed, and conveyed without loss of power from the powered control system either because they are very low-powered signals, or because external power is utilized.

There are several ways modulation signals, if they exist, can be highlighted on bond graphs. For example, in Fig. 5.12 it is intended that the control valve be manually operated such that its resistance Rv is Xv dependent. This can be illustrated as

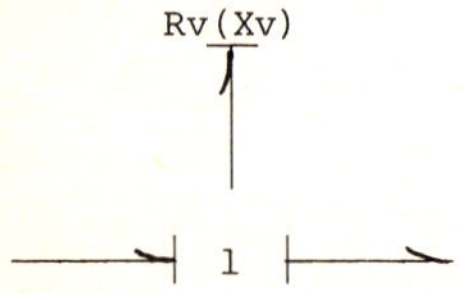

or as

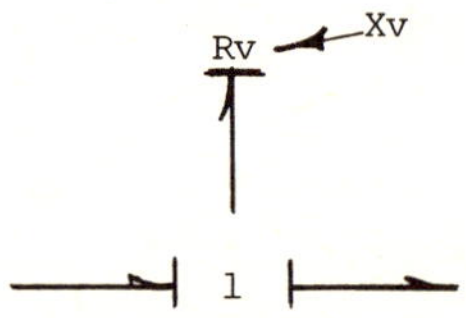

If modulation derives from external sources, it is convenient to use the left-hand representation. If modulation derives from feedback within the system, the right-hand method allows its physical origins to be illustrated.

For example, consider a pilot-operated pressure relief valve where the relief valve is situated at the pump but its operating pressure is sensed well downstream at the actuator. Fig. 5.13a illustrates this situation, and includes the pilot actuation line as a broken line. Fig. 5.13b shows the corresponding part of a bond graph representation of the system, and includes the modulating feedback effect as a broken line with a full arrowhead on it. Thus, the modulation bond is distinguished from power bonds by being a broken line of light construction, and by having a full arrowhead on it in the direction of signal flow. Note that the modulation bond, referred to in bond graph terminology as *an active or activated bond*, has only one power variable associated with it, in this case pressure Pa . The implication is that the other power variable is zero and therefore that no power flows in the active bond.

5.6.2 Modulated Transformers

The modulus of a TF or GY effect sometimes may be variable. The variability

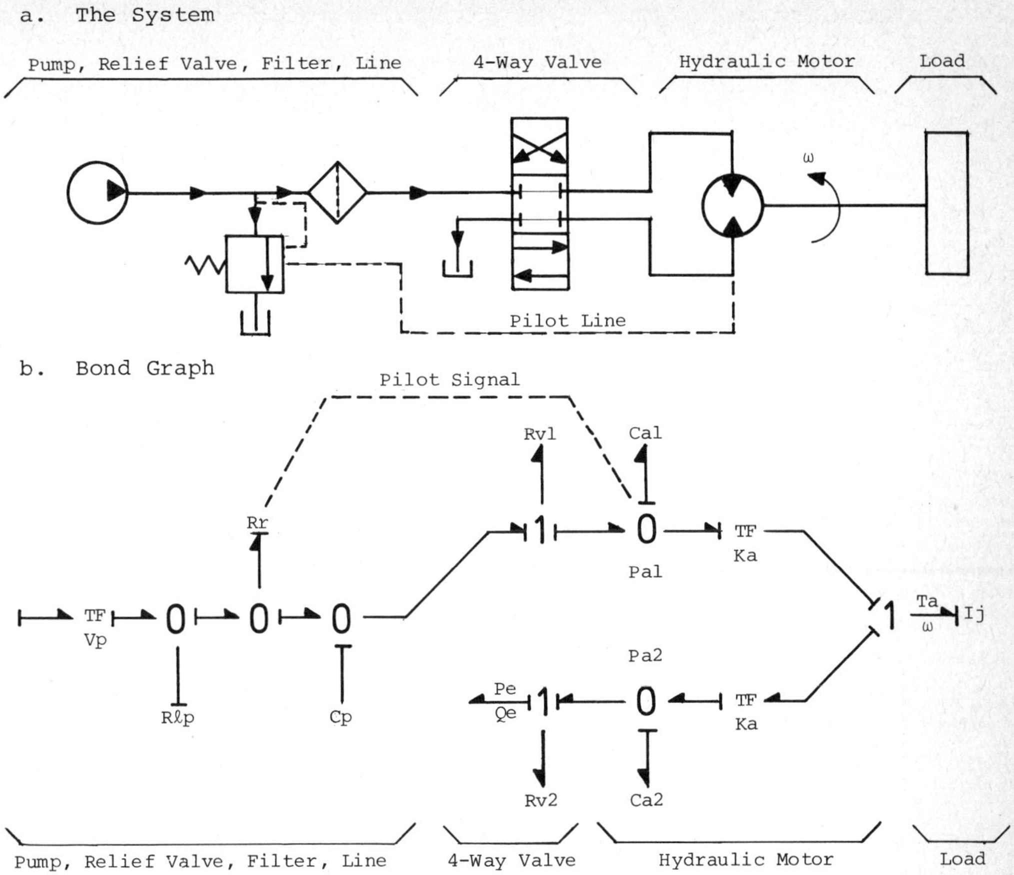

FIG. 5.13 SHOWING FEEDBACK ON A BOND GRAPH

Rr represents the Relief Valve.
Ca1 represents the inlet side of the Hydraulic Motor.
(The components of the bond graph will be explained in Ch. 7).

may be deliberately controllable by an external input, or it may be controlled by system feedback. A variable delivery pump is an example. TF or GY elements in this category are called *modulated transformers* and can be represented on bond graphs as MTF or MGY . Further, an active bond can be shown to indicate the source of the modulus variation. For example

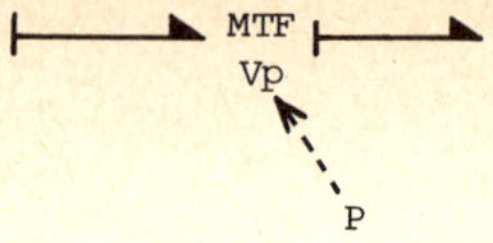

illustrates how a pressure-compensated pump could be indicated. The symbol illustrates simply that pump displacement per radian Vp is a function of pressure P , which could be a locally set or a fed-back pressure.

5.6.3 Fields and Junction Structures

There is a more sophisticated approach to bond graphs involving *fields* of C , I , and R effects, with *junction structures* of 0-junctions, 1-junctions, TF's and GY's to link the fields. Fields are multiport generalizations which allow concise representation on bond graphs of complex assemblages of C's , I's , and R's . There are situations involving dynamically complex mechanical motions or linkages, or complex electrical circuits, in which the field and junction structure concept is advantageous. However, it appears that the direct effect-by-effect approach to bond graphs is quite appropriate to the modelling of most hydraulic control systems. It is left to the reader to take up study of fields and junction structures via Ref. 5.1 when he is ready or in need.

5.6.4 Simplifications

With a little usage, it will become apparent that adjacent similar junctions can be combined. For example, on previous Fig. 5.8, the Rf and Ia 1-junctions are adjacent and can be expressed

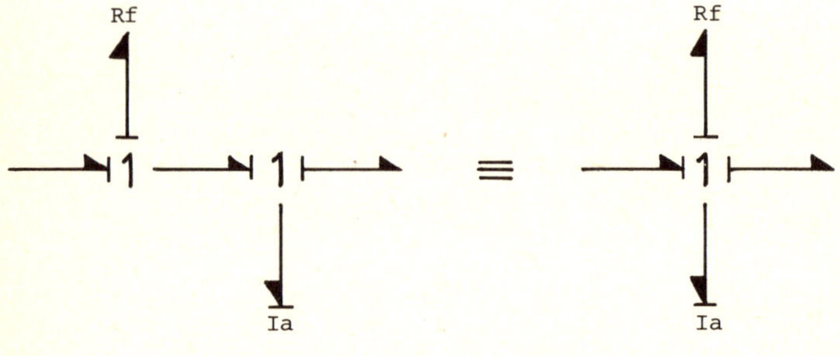

In the bond graphs of this Chapter, both variables of each bond were eventually added to the diagrams. At each junction, there is need to show the effort variable (0-junctions) or the flow variable (1-junctions) on only one of the bonds at the junction, as it is common on all bonds at the junction. It is quite appropriate to show the common variable against the junction itself, as follows

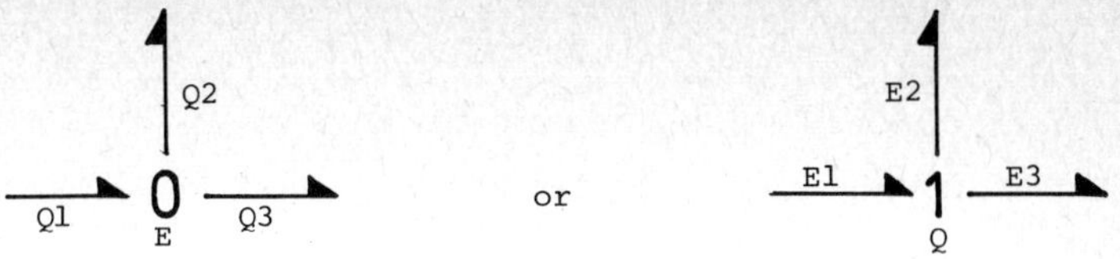

5.7 CONCLUSION

Bond graph terms and symbols have been introduced and their uses demonstrated. Only those aspects of bond graph techniques which have been found to be relevant to the present Text have been discussed in detail. For further reading in bond graph techniques and their uses in a wide variety of dynamic systems, Refs. 5.1 to 5.4 are recommended.

It is recommended that the user be systematic when putting power state variables on bonds. In the present Text, the effort variable is written above horizontally drawn bonds, and to the left of vertical bonds. The flow variables are written below the horizontal bond and to the right of the vertical bond.

It is recommended also that, initially at least, readers adopt the formal approaches to assigning causality and to preparation of equation sets given in Sections 5.4 and 5.5 respectively.

Chapter 6 gives a brief review of digital simulation, the model solution method most appropriate to bond graph models. Chapter 7 introduces in some detail the equation forms which might be required for use with the bond graph of a hydraulic control system. Chapter 8 demonstrates the formation and use of bond graph models for a variety of hydraulic systems, including electrohydraulic servo-systems with feedback shown as active bonds.

5.8 REFERENCES

5.1 Karnopp, D. and Rosenberg, R., System Dynamics; a Unified Approach, Wiley, 1975.

5.2 Thoma, J., Introduction to Bond Graphs and their Applications, Pergamon, 1975.

5.3 Transactions ASME J. of Dyn. Syst. Meas. and Control, V.94, Series G, N.3, 1972 (whole issue).

5.4 Journal of the Franklin Institute, V.308, N.3, 1979 (whole issue).

CHAPTER 6

SOLUTION OF POWER FLOW MODELS

6.1 INTRODUCTION

Dynamic mathematical models of physical engineering systems of the kinds relevant to the present Text, are formed in order to be used to predict a system's dynamic response, or to provide dynamic performance information. It is intended that the model be solved, fully or partially, to provide this information. When choosing a modelling procedure, the analyst needs to be aware of the solution procedures available to him. If, as in the not too distant past, only manual solution procedures are available, simplified models amenable to hand solution must be sought. If an analogue computer is to be utilized, a more complete model can be sought. However, there will be limitations on the number and types of non-linear expressions which can be included in the model. Digital computers provide the potential for solution of fully detailed dynamic models. In particular, the digital simulation procedures developed for use with most of the major computers provide a general solution procedure for a very wide range of dynamic models.

Power flow modelling procedures provide the system model as a set of low-order algebraic and differential equations. A detailed model of a system of some substance could be a set containing a substantial number of such equations. Digital simulation procedures are most compatible with this situation. There is no need —— in fact it is undesirable —— to run the low-order equation groups together to form more complex groupings of variables and coefficients. In the system design situation, in which the sensitivity of system performance to changes of some parameters is likely to be investigated, the simple presentation of coefficients and variables resulting from power flow modelling is positively an advantage.

Thus, power flow modelling and digital simulation are highly compatible. The existence of the digital simulation solution procedure has made it possible to realize the advantages of the power flow modelling procedure. Digital simulation is the only solution method considered to be relevant to the present Text.

Digital simulation will be dealt with only superficially. It is required here only as a user-orientated model solution procedure particularly suited to the purposes of the present Text. Each major computer is likely to have its own digital simulation procedure(s) or variations. Experience has shown that it is relatively easy to learn to utilize them. Though internally complex, the procedures are designed to be utilized by the user in a "black box" mode. The user has only to learn how to get his data, equations, and instructions into the "black box" solution procedure. In fact, the basic purpose of a digital simulation procedure is to bridge the user to the computer so that the user's simple statements can trigger and control the computer's complex and comprehensive problem-solving capability.

A number of relevant references are listed at the end of Chapter 6.

6.2 DIGITAL SIMULATION

The digital simulation procedures relevant to the present Text are generally classified as Continuous System Simulation Languages (CSSL's). A continuous system is regarded as one which can have a continuous time history associated with its dynamic performance, rather than one intended or known to have discrete states in time.

Fig. 6.1 illustrates the relationship of the CSSL's among the dynamic problem solving languages commonly used in digital computing. To help orientation of the reader, some of the specific languages associated with each grouping are included. The high level problem solving languages such as Fortran can be utilized directly to prepare a model for solution. However, the CSSL's (for continuous systems) and the discrete system languages provide the bridge between the user and the computer's basic compiling language (Fortran or alternative).

The CSSL's fall into two categories. Those in the block-orientated group are intended to fairly directly rival analogue computing, in which a block diagram type of model is most appropriate. The expression-orientated CSSL's anticipate receiving the model in the form of a set of equations. It is this group which is of primary interest in the present Text, and only this group will be considered further.

Characteristics desirable in a CSSL include:

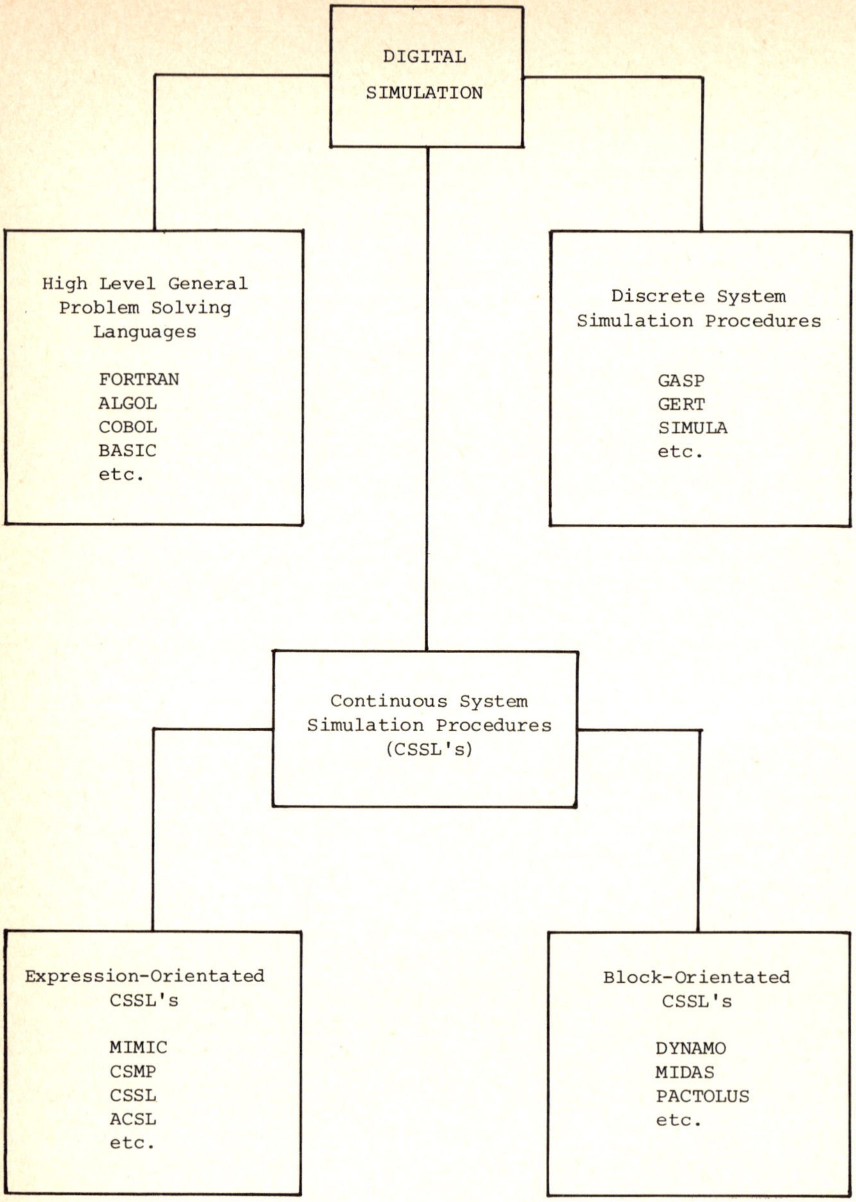

FIG. 6.1 APPROACHES TO DIGITAL SIMULATION

- easy utilization, with little computing expertise required of the user;
- good man-machine rapport;
- good result display capability;
- economy in computer use;
- speed, in both utilization and computing;
- accuracy: numerical instability is a potential problem. To avoid it requires choice of an integration procedure, and integration step size, compatible with the particular problem being solved. For efficient use of the computer, multiple rate is preferred. This allows integration step size to vary from integration variable to integration variable, and even during a particular integration. A slowly changing variable does not require the small integration step size necessary with rapidly changing variables;
- parallel statement of model equations: this allows model sections or equations to be worked out and specified in any order. The computer does the sorting into its execution (i.e. procedural) order. Problem languages such as Fortran require procedural programming, i.e. the computer executes the equations and statements in the order they are supplied. Though parallel programming is an illusion —— the computer must convert the statements received by it into procedural form before execution starts —— it is real as far as the programmer is concerned. Parallel statement specification is much preferred to the procedural form for most sections of a digital simulation;
- the ability to specify sections of a model in algebraic form, with separate specification of numerical values for coefficients and parameters. This macro capability is particularly desirable in system design, when the effects of value changes need to be studied. It is also very useful where response optimisation is required;
- the ability generally to incorporate common operations, functions, and signals as "building blocks" which can be called into required positions in a system being simulated. In general, this macro capability allows access to stored "library" sub-routines (preferably in either Fortran or the CSSL language). It should also include a capability for logic control of statement execution —— this is needed to represent some common non-linear relationships.

6.3 EXPRESSION-ORIENTATED CSSL's

The CSSL's date back to the mid-1950's. Ref. 6.1 contains a survey of their early developments, which were of considerable diversity. In 1967 the Simulation Councils Incoporated (now the Society for Computer Simulation) published specifications for standardizing CSSL development. Ref. 6.2 contains details. The most recent CSSL's are close to but not precisely in line with the specifications. The specifications are themselves ambiguous or incomplete in parts. Improved accessing methods and new requirements complicate the search for user-advantaging standardization of CSSL's. Ref. 6.3 gives information on developments to the mid-1970's, and Ref. 6.10 gives a hydraulic control system's orientated review of the situation to 1979.

MIMIC (Control Data Corporation) and DSL/90 (IBM) were perhaps the major links drawing the early digital simulations towards the standard CSSL concept. Both were widely and generally used. Control Data Corporation, influenced by the standardization move, now has CSSL 3 [6.4]. Similarly, IBM's DSL/90 has been developed into CSMP 3 (Continuous System Modelling Program), [6.5]. Xerox Data Systems SL-1 [6.6] is particularly close to the standard CSSL. Mitchell and Gauthier [6.7] have produced ACSL, pronounced "axle" and referring to Advanced Continuous Simulation Language, for use with most computers. Most CSSL's have been produced for use with a particular company's computers. Thus, ACSL represents a major step forward.

Apart from the general purpose CSSL's, there exist many special purpose orientated digital simulation procedures. For example, HYDSIM [6.8] is especially designed for hydraulic control system response analysis using power port models of components. ENPORT [6.9] has been especially developed to accept a model in bond graph form. It has the disadvantage (very serious in the present Text) of being able to handle only linear expressions. Both HYDSIM and ENPORT have the significant feature of being able to store "standard" models of components likely to be used and re-used in various control systems. Some other CSSL's are orientated to incorporate response optimization schemes.

6.4 CONCLUSION

CSSL procedures provide the designer-analyst of engineering control systems with the model solution procedure closest to his needs. They can have the solution speed, accuracy, man-to-system rapport, and the display and plotting characteristics of analogue computers. Their computational capability far outstrips that of analogue computers. So too does their general availability. Increasing availability of interactive access and of minicomputers allows the "hands on" access previously the preserve of analogue computers.

Digital simulation is not necessarily a panacea for solving dynamic models.

Programming and running time can be costly if inexpertly used. Round-off error and precise accuracy may be difficult to estimate. Numerical instability can be a problem where different parts of a model have widely different response speeds. This situation may be reflected by the presence in the model of widely different natural frequencies. These kinds of model are referred to as stiff. Numerical stability in the solution of stiff models is being better overcome with the development of more flexible integration procedures which do not necessarily require formal explicitness of all of the relationships of the model. ACSL edges into this category. It may require some experience in particular problem areas before the best integration procedure and step size specification can be made with confidence. A modern CSSL may have as options several Runge-Kutta integration procedures plus several predictor-corrector types (i.e. Milne). Some are more complex than others, requiring more computer capability. Thus a simple self-starting Runge-Kutta 4th order procedure should be preferred to the more complex Milne 5th order form (needs starter, complex but very accurate) unless it is known to be inadequate to a particular set of equations or relationship.

CSSL's are very compatible with power flow models. They are also very compatible with the equation sets and relationships which are relevant to hydraulic control system response, including response optimization procedures.

Results of completed simulations of bond graph models using digital simulation are included in Chapter 8.

6.5 REFERENCES

6.1 Branden, D., Burwen, M., and Wachouski, A., Digital Simulation of Physical Systems, Assoc. for Computing Machinery.

6.2 Strauss, J.C., The SCI Continuous System Simulation Language (CSSL), Simulation, V.9, N.6, Dec. 1967, pp.281-303.

6.3 Benyon, P.R., Improving and Standardizing Continuous Simulation Languages, Proc. SIMSIG Simulation Conf., Melbourne, Australia, 1976.

6.4 Continuous System Simulation Language 3 (CSSL 3) Version 1, Control Data Corp., Sunnyvale, California, 1974.

6.5 Continuous System Modelling Program 3 (CSMP 3) and Graphic Feature, IBM Canada, Ontario, Canada, 1972.

6.6 SL-1 Reference Manual, Xerox Data Systems, El Sequndo, California, 1970.

6.7 Advanced Continuous Simulation Language (ACSL) User Guide/Reference Manual, Mitchell and Gauthier Assoc., Concord, Mass., 1975.

6.8 Smith, C.K., HYDSIM User's Manual, School of Engineering, Oklahoma State University, 1973.

6.9 Rosenberg, R., A Users Guide to ENPORT 4, Wiley Interscience, 1975.

6.10 Iyengar, S.K.R., Review of Computer Software for Simulating Hydraulic Systems, BFPR Journal, V.12, N.4, pp.323-329, Fluid Power Research Centre, Oklahoma State University, 1979.

6.11 Benyon, P.R., Some Recent International Activities in Simulation, Proc. 4th Biennial Conf. Sim. Soc. of Aust., University of Queensland, Australia, 1980.

CHAPTER 7

SELECTING EQUATIONS AND COEFFICIENTS

7.1 INTRODUCTION

The bond graph material of the Text to here has emphasized formation of the structure of a system's dynamic model rather than the actual equations whose solution will provide dynamic response. Equations have been presented in some examples, but with little discussion.

It is valid to emphasize the model structure. To the experienced bond graph analyst the model structure is the model. He "sees" his physical system in the bond graph structure. In a particular technology —— in the present case hydraulic control systems —— the physical equations used are of relatively few forms and these appear over and over again in the models of different systems within that technology. Thus the bond graph analyst also "sees" the equations implicit on the structure. To be sure, there is often a choice to be made on how detailed a description (i.e. an equation) of a particular action should be in a particular analysis. The representation of friction is one case. The representation of valve flowrate versus pressure drop and valve displacement is another.

However, specific equations for each S , R , C , I , TF and GY on a bond graph do have to be specified before simulation can progress. Further, when all equations have been selected and written down, there remains the selection of numerical values for the various coefficients in the equations. For a particular system whose hardware has been provisionally specified, many of the coefficients required for simulation of the dynamic model can be obtained as design specifications or from component manufacturer's data. However, there is usually a small number of coefficients whose values are less easily determined.

The present Chapter brings together physical relationships which are relevant to the Text. Where ambiguity or choice exists, some discussion is offered. Selection of values for those coefficients which are not straightforward is discussed, and some values for coefficients are included.

7.2 COMPLIANCE

7.2.1 Introduction

No liquid is fully incompressible. No mechanical connection is completely rigid. Both retain elasticity characteristics which can be likened to the characteristics of stiff springs. The compliance characteristic of the oil in a hydraulic control system is a vital parameter affecting system dynamic response. Mechanical compliance in the various mechanical components of the system may affect system response, depending on the degree of compliance present. Hydraulic lines —— particularly hoses —— can expand radially under pressure, mechanical shafts can suffer torsional strains, actuator rods suffer compression or elongation, beams suffer deflection, and so on. Because of its significance to system performance, oil compliance will be considered in some detail. Mechanical compliance, being better understood by most engineers, will be dealt with briefly only.

7.2.2 Oil compliance

Oil compliance is a key characteristic in the action and dynamic performance of hydraulic control systems. Initially it might be considered desirable that the oil has a very low compliance (i.e. very high resistance to compression), an incompressible liquid being the ideal. Reflection indicates that this is not the case. An incompressible liquid subjected to positive pumping action which is opposed by a significant downstream impedance, as is the case with most hydraulic control systems, would experience instant and extreme high pressure development. This would impose intolerable strain on the mechanical components containing the oil (pump lines, filters, valves, actuators). It takes time for even the fastest-acting pressure relief valves to respond. On the other hand, it is obvious enough that a highly compressible (i.e. high compliance) fluid is incompatible with fast response, as the stiff coupling between control and load necessary for fast response is lost. A pneumatic servo can never be as stiff and as fast in response as its hydraulic counterpart.

Fig. 7.1 illustrates the idea that for a particular hydraulic system and its driven load, the pressure development at the actuator following a control command at the control valve might look like

. curve 1 if the oil was virtually incompressible
(i.e. very low compliance, or very high bulk modulus)

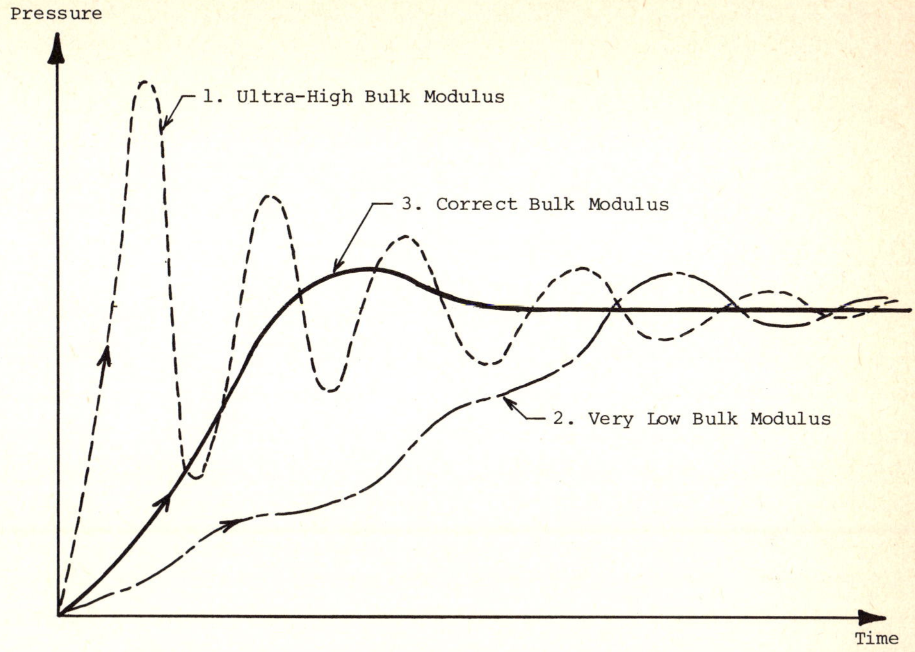

FIG. 7.1 ILLUSTRATING THE EFFECT WHICH BULK MODULUS OF FLUID CAN HAVE ON THE PRESSURE DEVELOPMENT IN A HYDRAULIC CONTROL SYSTEM

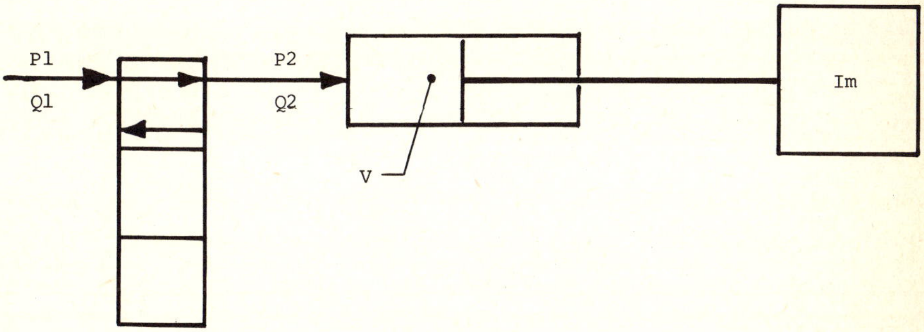

FIG. 7.2 FLOW THROUGH VALVE INTO ACTUATOR

. curve 2 if the oil was quite compressible
(i.e. high compliance, or very low bulk modulus)

. curve 3 if the oil compliance was about right.

Thus, the actual value of bulk modulus for a hydraulic system liquid is important. It should not be so high that excessive pressures are built up, nor so low that stiff direct control of the load is lost. It seems that light mineral oils, with a natural quiescent bulk modulus of around 1750 MPa are about right. It is not clear whether this is close to a natural optimum figure, or whether hydraulic components and systems have evolved recognizing that mineral oils were their natural fluids.

The compliance characteristic of a liquid can be described by a bulk modulus defined by the static equation

$$B = - V \cdot \Delta P/\Delta V \qquad 7.1(a)$$

where V is the initial volume of liquid subjected to compression
ΔV is the change in volume V due to compression
ΔP is the change in pressure which causes compression ΔV .

If volume V of liquid is subjected to *increased* pressure ΔP , V *decreases* by ΔV , i.e. $\Delta P/\Delta V$ is negative, and the negative sign of equation 7.1(a) is used to make bulk modulus B a positive value.

For dynamic performance analysis, the need is to have a relationship which enables calculation of pressure development in chambers into which hydraulic fluid is flowing. Fig. 7.2 shows a typical case —— flowrate $Q1$ at pressure $P1$ upstream of the open control valve becomes $Q2$ at $P2$ as it flows into the actuator. Neglecting leakage losses, transient $Q2$ will be less than $Q1$ by an amount Qc known as the compressibility component of the incoming flowrate. That is, $Q2 = Q1 - Qc$ describes the dynamic flows following activation of the control valve. In steady-state $Q1 = Q2$, as the effect on the specific volume of the fluid of the low steady-state pressure difference can be neglected.

Equation 7.1(a) can be expressed

$$B = - V \cdot dP/dV \qquad 7.1(b)$$

This equation can be made dynamic by using dP/dt for dP and dV/dt for dV , giving

$$B = - V \cdot (dP/dt)/(dV/dt) = - V \cdot \dot{P}/\dot{V} \qquad 7.1(c)$$

Now for the situation of Fig. 7.2, $\dot{V} = -Qc$, and

$$B = V \cdot \dot{P}/Qc \qquad 7.1(d)$$

Then,

$$P = B/V \cdot \int Qc\ dt + P(o) \qquad 7.1(e)$$

Equation 7.1(e) is readily seen to be a linear capacitance type relationship, with capacitance defined as

$$C = V/B \qquad 7.1(f)$$

Equation 7.1(a) can be expressed

$$\Delta P = -\frac{B}{V} \cdot \Delta V \qquad 7.1(g)$$

For a particular B , V and ΔV , ΔP will be a particular value. For the same B and ΔV , but an increased volume V , ΔP will be smaller. That is, pressure development is dependent on the magnitude of the volume V under compression or expansion. It is important to keep this in mind when utilizing equation 7.1(e) for systems involving linear actuators. As oil flows into the actuator, displacing its piston, the volume of oil under compression changes. This is readily accounted for by expressing the volume as

$$V = V(o) + Xa \cdot Aa \qquad 7.2$$

where $V(o)$ is a suitable initial or mean value for the volume of oil under compression

Xa is piston displacement from the position $V = V(o)$

Aa is the effective area of the actuator piston.

Note that for single rod actuators, Aa is one value for the head side and another (smaller) value for the rod side. The problem of varying volume does not arise with rotary drives, the volume of oil under compression on either side of the rotary actuator remaining virtually constant.

Oil compliance may have to be considered in any situation where oil discharging into or through a volume is thought to have a significant effect on system response by affecting pressure development. Possible situations of this type include a pump discharging into a line, flow into and through filters, flow through lines, flow into actuators, flow from actuators through lines and possibly filters of significant fluid capacitance. Whether or not the fluid capacitance effect of a particular component should be considered in the dynamic analyses of a system is largely a matter of experience and judgement.

7.2.3 Values for Bulk Modulus

The bulk modulus of a particular liquid is affected by its temperature and its pressure, though the effects are not significant in the working pressure and temperature ranges associated normally with hydraulic control systems fluids. Aeration of the fluid, prevalent in working hydraulic systems, is a much more significant factor. A typical mineral hydraulic fluid in its quiescent state has a bulk modulus of around 1750 MPa. Under the influence of turbulence and cavitation in working hydraulic systems, bulk modulus might drop locally to 10% of this value, or even to zero locally if cavitation is fully developed. The lack of drive stiffness caused by low bulk modulus can seriously affect the response and performance of hydraulic control systems. In its cycle through a particular system the oil's bulk modulus can vary from a high value in the reservoir, to a lower value after pumping, to a lower value again after expansion through the control valve, and so on. In a well designed reservoir, the oil has a chance to regain or approach its quiescent bulk modulus before again being drawn into the working system. Merritt [7.1] discusses and provides typical values for bulk modulus at different parts of working hydraulic systems. A detailed discussion of bulk modulus and the variations in its definition is provided in Ref. 7.2.

7.2.4 Mechanical Compliance

Hydraulic control systems include numerous mechanical components, each of which has compliance in the form of elasticity. For example

- the shaft between drive and pump is subjected to torsional deflection when under torque,
- the hydraulic lines (particularly hoses) can expand under pressure,
- the rod of a loaded linear actuator is under compression or tension. It may also be subjected to a degree of column buckling effect,
- linkages used to connect actuator to load, prevalent in mobile lifting and digging equipment for example, are subjected to deflection when under load.

Most hydraulic systems are designed such that the mechanical components are very stiff and will experience minimal deflection under load. Therefore, mechanical compliance

- can often be neglected when predicting the response of a system,
- if significant, is usually well within the elastic range of the component, allowing use of well-established mechanical load-deflection relationships.

If it is required, mechanical compliance usually can be represented as a Ck effect from an 0-junction, as shown on Fig. 7.3. It is usual also that the Ck effect can be taken as linear, in which case Fig. 7.3 can be interpreted as

$$\Delta\dot{X} = \dot{X}1 - \dot{X}2 \qquad 7.3$$

$$F = 1/Ck \cdot \int \Delta\dot{X}\, dt + F(o) \qquad 7.4$$

$$= 1/Ck \cdot \Delta X + F(o) = Kk \cdot \Delta X + F(o) \qquad 7.4(a)$$

where $\dot{X}1$, $\dot{X}2$ are the instant velocities of the two sections of the compliance being considered

$\Delta\dot{X}$ is the velocity difference between $\dot{X}1$ and $\dot{X}2$

Ck is the compliance of the elastic mechanical component

Kk = 1/Ck is the "stiffness" of the elastic mechanical component

F is the force developed in and transmitted by the compliance

ΔX is the instant deflection between the two sections considered.

Note that equation 7.4 is in the form required for power flow modelling whereas equation 7.4(a) is the conventional static form for the force-deflection relationship for a linear spring.

Equations 7.3, 7.4, 7.4(a) have the equivalents for rotation motion

$$\Delta\omega = \omega 1 - \omega 2 \qquad 7.5$$

$$T = 1/Ck \cdot \int \Delta\omega\, dt + T(o) \qquad 7.6$$

$$= 1/Ck \cdot \Delta\theta + T(o) = Kk \cdot \Delta\theta + T(o) \qquad 7.6(a)$$

where $\omega 1$, $\omega 2$ are the instant angular velocities of the two sections of the compliance being considered

ω is the difference between $\omega 1$ and $\omega 2$

T is the torque developed in and transmitted by the compliant component

Ck is the compliance coefficient of the component

Kk = 1/Ck is the "stiffness" coefficient of the component

$\theta = \int \Delta\omega\, dt$ is the angular deflection involved.

If appropriate, non-linear Ck relationships can be used. For example, for the load-deflection characteristics of non-uniform cross-section beams.

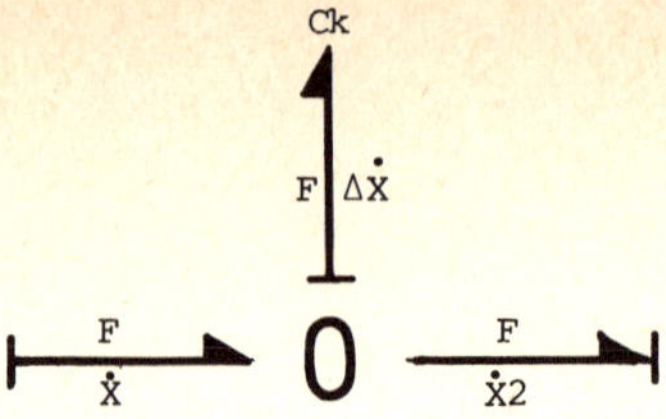

FIG. 7.3 REPRESENTING MECHANICAL COMPLIANCE

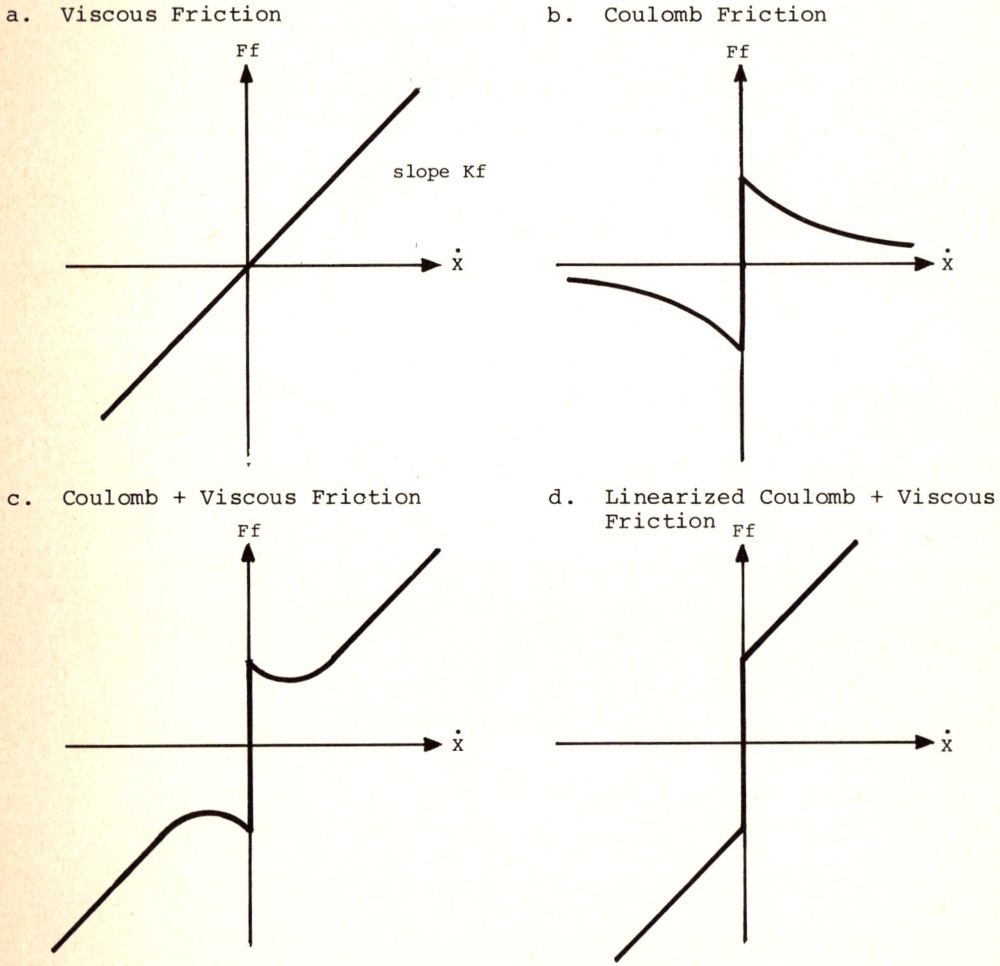

FIG. 7.4 REPRESENTING FRICTION

7.3 FRICTION

Friction is present wherever one mechanical part moves on and relative to another. There are many such situations within working hydraulic systems. For example, the pistons moving in their bores in a piston pump or motor; the piston head in a linear actuator; the piston rod moving through the gland of a linear actuator; etc. The most significant friction, however, is likely to be associated with the load driven by the system. Substantial loads require the support of heavy bearings and guides.

Friction in these various circumstances can range between the two limits:

1. *Viscous friction*, where good film lubrication is implied. That is, the mating mechanical parts are separated by a film of lubricant, so that only fluid shear is present to resist motion. Under these circumstances

$$Ff \propto \dot{X} \qquad 7.7(a)$$

$$Ff = Kf \cdot \dot{X} \qquad 7.7$$

where Ff is the force required to overcome friction
$\dot{X}$ is the relative velocity between the moving parts
Kf is the viscous friction coefficient

provides an accurate description of friction force for use in dynamic simulation. Fig. 7.4a illustrates the $Ff - \dot{X}$ relationship of equation 7.7.

2. *Coulomb friction*, where the mating mechanical components are in direct contact, with incomplete or no effective lubrication. In this situation static friction has to be overcome before relative motion can start, and the static friction force $F\mu$ is described by

$$F\mu = \mu \cdot Fn \qquad 7.8$$

where μ is the coefficient of friction existing between the components
$F\mu$ is the force required to start motion
Fn is the normal force existing between the components.

It is well known that when coulomb friction is overcome and motion starts, the friction force drops sharply. Fig. 7.4b illustrates coulomb friction.

Many of the components in a hydraulic control system are flooded with the hydraulic fluid. One of the properties desired of a hydraulic fluid is that it be a good lubricant. This is certainly realized if a mineral oil is used, as is the case for most hydraulic systems. Thus, good lubrication can be expected for the internal moving parts of pumps, motors, actuators, and valves. However, the presence of generous quantities of oil does not absolutely guarantee that no direct contact of parts takes place. This is particularly so if the parts have stationary periods, such as is usual with linear actuators, spool valves, and hydraulic motors. A film of lubricant developed and stabilized when parts are in motion can be squeezed out to some degree when motion stops. It is preferable to allow for this effect when considering friction of even well-lubricated parts, by using a friction force-velocity characteristic of the type illustrated on Fig. 7.4c.

Fig. 7.4d illustrates a reasonable idealization of friction in well lubricated components. If the dry friction band is narrow, it can be neglected in a system simulation, leaving frictional force as viscous (equation 7.7). If the "stiction" effect is considered to be significant friction can be idealized to the form

$$Ff = F\mu \cdot \mathrm{sgn}\dot{X} + Kf \cdot \dot{X} \qquad 7.9$$

where sgn means "sign of", allowing that $\dot{X}$ can be positive or negative.

In extreme cases a characteristic of the form of Fig. 7.4c can be described by equation or numerically. However, the overriding requirement for successful dynamic analysis and simulation should be kept in mind, i.e. the simplest model adequate for the proposed simulation is the best! It is surprising how often that an assumption that friction in hydraulic system components is approximately viscous proves adequate for hydraulic control system performance predictions. Of course, the more suspect the lubrication present the more suspect is a viscous friction assumption. However, it is reasonable that a system designer can assume that the moving parts of his proposed system components, and of the load being driven, are reasonably lubricated.

Frictional forces and torques always oppose motion and irreversibly dissipate (waste) power. However, it should not be presumed that friction should be eliminated or even minimized. Friction is often the only mechanism providing damping of load response to sharp inputs. A minimum-friction actuator driving a minimum-friction load may well prove too oscillatory, or even result in limit cycling which borders on instability.

Numerical values for friction coefficients are included in some of the examples of Chapter 8.

7.4 MODELLING DRIVEN LOADS

7.4.1 Inherent Loads

The positioning of inertia loads in the presence of friction, possibly of structural compliance, and possibly of external load forces, is the task faced by the majority of hydraulic control systems. The designer-analyst needs to understand the basic load arrangements he may have to choose between or adapt from. For example to control the rudder of an aircraft, the control system needs to overcome inertia, friction, structural compliance, and the aerodynamic forces on the rudder. The hydraulic system of a large press has to cope with inertia of the press head assembly, friction in the press guides and bearing, deflections in the press frame, and the large load induced when the press head comes in contact with the work piece.

Fig. 7.5 illustrates eight basic load arrangement models, together with their bond graphs, causal requirements, and probable (not all exclusive) equations. The first three entries are straightforward. The force Fa developed in the actuator is the causal input variable, load-actuator velocity is the resulting effect. There is no compliance effect between actuator and load, and this is often a reasonable assumption.

Entry 4 allows for structural compliance in the drive or mountings between actuator and load. The C effect causes actuator velocity $\dot{X}a$ and load velocity $\dot{X}m$ to be different, by the amount $\Delta\dot{X}$ shown on the bond graph. The preferred causalities for Ck and for Im both require the causal bar to be on the left end of the $Fa \cdot \dot{X}a$ bond. This is a significant variation from the load input bond causalities of Entries 1 to 3, all of which implied stiff (non-compliant) connection between actuator and load. The bond graph of Entry 4 is telling us that force Fa is a back effect for the Entry 4 drive, and that actuator velocity $\dot{X}a$ is the cause effect. This is a most significant revelation, with strong implications as to how the actuator model (bond graph) must be presented for connection to a compliant drive.

Entry 5 allows for load friction in the compliant drive model.

Entry 6 illustrates a compliant drive between actuator and load, with damping or friction also present in the connection. This might be the situation if actuator friction was considered to be significant or even necessary to damp down oscillations in Xa and Xm due to the presence of compliance Ck . Again, and for the same reasons as for Entry 4, actuator force Fa is a back effect resulting from causal velocity $\dot{X}a$. The 0-junction on the bond graphs tells us that the compliant drive causes $\dot{X}m$ and $\dot{X}a$ to be different by the amount $\Delta\dot{X}$ (and consequently Xm and Xa to be different by ΔX). The 1-junction allows that both Ck and Rf affect $\Delta\dot{X}$, and that they act in parallel with each other.

Entry 7 adds the external force Fi to the situation of Entry 6. Fi is not necessarily constant, nor necessarily linear, and is a source.

Entry 8 shows the situation of Entry 3 operating in the vertical plane, when gravitational force acting on the mass must be included. This is shown as the constant force SFg (Source, Force, gravitational). SFg could have been added to the single 1-junction of Entry 3.

7.4.2 External Load Forces

Most hydraulic control system loads include inertias. Many also include other forces. For example, a machine tool drive or tool positioner will experience cutting forces. The tail-plane control on an aircraft will experience an aerodynamic force which is dependent on both the speed of the airflow over the surface and on the angle of attack of the surface to that airflow. A sensitive model for an electrohydraulic servovalve may have to account for both the inertia of the valve spool and the flow-forces generated on it.

Consider the case of an aircraft's rudder control. Both the inertia of the rudder, and the aerodynamic forces on it should be considered when assessing its position response to input commands. Whatever else may have been included in a bond graph for the control system and its load, the structure should end with the inclusion of an Rθ , vel bond from a 1-junction and with causal bar at the junction, indicating recognition of a force which is dependent on rudder angle (θ) and on airspeed (vel) . It is of no consequence whether Im is shown projecting from the same 1-junction. It can project horizontally as previously, or at any other angle for that matter.

7.5 LEAKAGE FLOWRATES

7.5.1 Relationships

The working parts of pumps, valves, and actuators must have clearances. For example, the pistons in their bores in piston pumps and actuators, the spool in its bore in spool valves, the gears and the side plates in gear pumps, etc. These clearances, though very small, will act as leakage paths when under high pressure drops. Normal hydraulic fluids are sufficiently viscous, and working clearances are sufficiently small, that leakage flows can be assumed to be laminar. That is

$$Q\ell \propto \Delta P \qquad 7.10(a)$$

where $Q\ell$ is the leakage flowrate through the leakage path
ΔP is the pressure drop which is inducing leakage.

Then,

$$Q\ell = 1/R\ell \cdot \Delta P \qquad 7.10$$

SYSTEM	BOND GRAPH	EQUATIONS	COMMENTS
1. Mass only 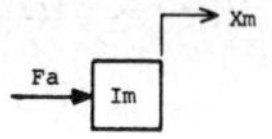	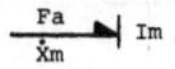	Fa = causal force $\dot{X}m = 1/Im \cdot \int Fa\, dt + \dot{X}m(o)$ $Xm = \int \dot{X}m\, dt + Xm(o)$	Fa derives from driving component. Newton's law. Auxiliary equation.
2. Mass with friction 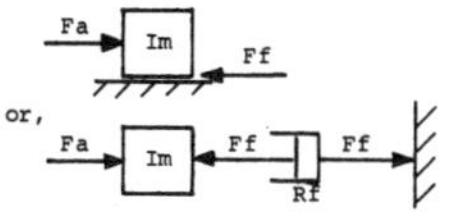	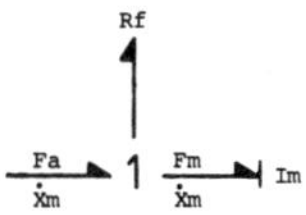	Fa = causal force $\dot{X}m = 1/Im \cdot \int Fm\, dt + \dot{X}m(o)$ $Ff = R(\dot{X}m)$ $= Kf \cdot \dot{X}m$ for visc.fric. $Fm = Fa - Ff$	R() denotes "a resistive function of". Kf is visc. fric. coeff. Net force on Im .
3. Mass with friction and compliant restraint 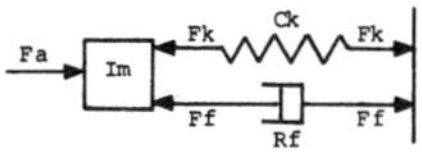	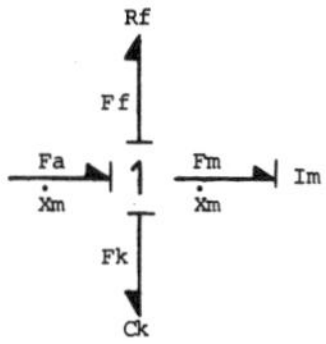	Fa = causal force $\dot{X}m = 1/Im \cdot \int Fm\, dt + \dot{X}m(o)$ $Ff = R(\dot{X}m)$ $Fk = C(\dot{X}m)$ $= Kk \cdot \int \dot{X}m\, dt + Fk(o)$ $Fm = Fa - Ff - Fk$	C() denotes "a capacitive function of". For linear spring, of coeff. Kk .
4. Mass, with compliant drive 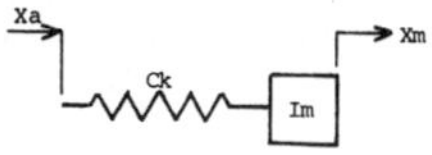	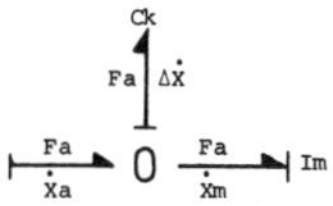	$\dot{X}a$ = causal input $\dot{X}m = 1/Im \cdot \int Fa\, dt + \dot{X}m(o)$ $\Delta\dot{X} = \dot{X}a - \dot{X}m$ $Fa = C(\Delta\dot{X})$ $= Kk \cdot \int \Delta\dot{X}\, dt + F(o)$	Fa cannot be the causal input, as development of spring deflection is required before F is developed. Thus, $\dot{X}a$ is the causal input for this configuration. F is a capacitive function of $\Delta\dot{X}$. For linear spring.
5. Mass, compliant drive, friction 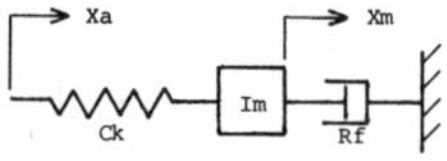	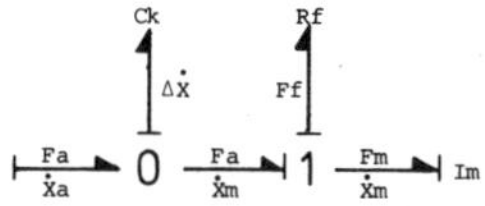	$\dot{X}a$ = causal input $\dot{X}m = 1/Im \cdot \int Fm\, dt + \dot{X}m(o)$ $\Delta\dot{X} = \dot{X}a - \dot{X}m$ $Fa = C(\Delta\dot{X})$ $Fm = Fa - Ff$ $Ff = R(\Delta\dot{X})$	As for 4. $\dot{X}a$ is caused by the component driving the load system. Use desired equation for friction.
6. Mass, with compliance and friction in the drive 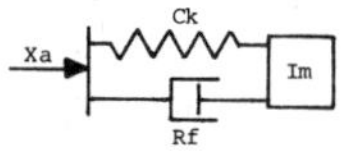	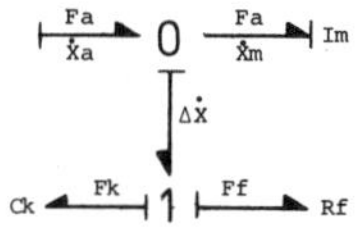	$\dot{X}a$ = causal input $\dot{X}m = 1/Im \cdot \int Fa\, dt + \dot{X}m(o)$ $\Delta\dot{X} = \dot{X}a - \dot{X}m$ $Ff = R(\Delta\dot{X})$ $Fk = C(\Delta\dot{X})$ $Fa = Ff + Fk$	As for 4 and 5, i.e. for compliant drives 4, 5, 6, velocity is the cause variable and force the resulting effect variable. Interesting, isn't it?
7. As for 6, plus external force Fi resisting motion 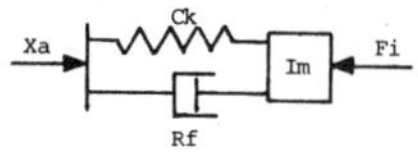	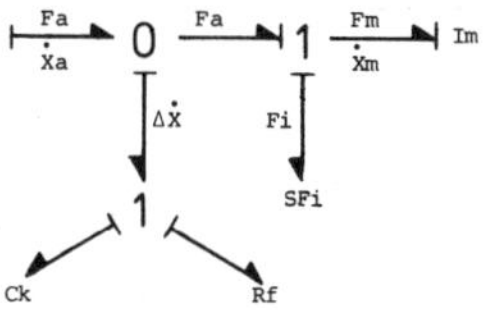	$\dot{X}a$ = causal input $\dot{X}m = 1/Im \cdot \int Fm\, dt + \dot{X}m(o)$ Fi = specified force $Ff = R(\Delta\dot{X})$ $Fk = C(\Delta\dot{X})$ $\Delta\dot{X} = \dot{X}a - \dot{X}m$ $Fm = Fa - Fi$	SFi indicates Fi is a source Allowing for effect of Fi .
8. As for 3, but suspended vertically 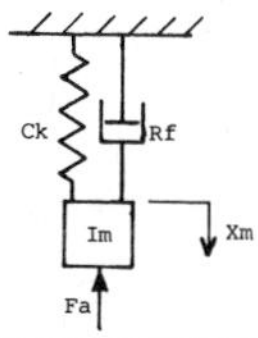	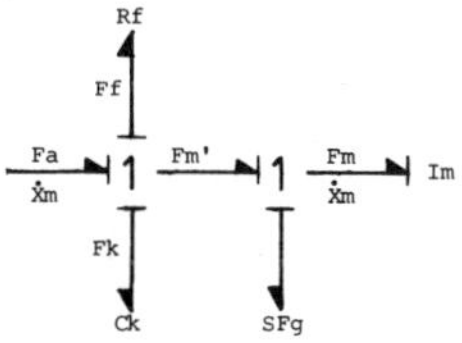	Fa, $\dot{X}m$, Ff, Fk: As for 3. $Fm' = Fa - Ff - Fk$ $Fm = Fm' - SFg$	SFg is constant force exerted on mass due to gravity. The two adjacent 1-junctions can be combined.

FIG. 7.5 INERTIA LOAD SYSTEMS, WITH VARIOUS FRICTION, COMPLIANCE, AND EXTERNAL FORCE CONSTRAINTS

where $R\ell$ is the resistance of the leakage path.

ΔP will be a variable in the power state equations and the simulation. Usually, leakage flows back to tank, the discharge side pressure can be taken as zero, and ΔP approximates the pressure at the entry to the clearance. Thus, leakage from a pump can be expressed

$$Q\ell p = 1/R\ell p \cdot Pp \qquad 7.10(b)$$

where Pp is the pressure at pump discharge

$R\ell p$ is the leakage resistance coefficient of the pump.

Analytical estimation of a particular $R\ell$ may be complex. For example, for a piston located coaxially in its bore, as illustrated on Fig. 7.6a, leakage is given by

$$Q\ell = \left[\pi \cdot D \cdot H^3/(96 \cdot \mu \cdot L)\right] \cdot \Delta P \qquad 7.11$$

where D is piston and bore diameter

H is the diametral clearance between piston and bore

μ is dynamic viscosity of the liquid

L is the length of the piston

ΔP is the axial pressure drop which is inducing leakage.

In this case, it is readily seen that for the single piston

$$R\ell = 96 \cdot \mu \cdot L/(\pi \cdot D \cdot H^3) \qquad 7.12$$

However, the attitude of a piston working in its bore can rarely be guaranteed. The piston can tilt, and can become eccentric in the bore as illustrated on Figs. 7.6b, c, respectively. For the case of full eccentricity, Fig. 7.6d, it is readily shown that $Q\ell$ will be 2.5 times the value calculated from equation 7.11. That is, for full eccentricity, $R\ell$ drops to 0.4 of its value for the concentric case of Fig. 7.6a. It is worth observing that leakage is proportional to clearance cubed. That is, if clearance is doubled, the resistance is decreased to an eighth, and the leakage flowrate goes up eight times. It is also worth noting that clearance is affected —— increased —— by pressure expansion effects, increasing leakage from the value calculated using manufactured clearance. Leakage in the clearances of piston components is discussed more fully in Ref. 7.3.

7.5.2 Coefficients

Empirical evaluation of an $R\ell$ coefficient is relatively simple. For a particular pump, for example, the overall $R\ell p$ (pistons, port plate, etc.) can be established by plotting measured $Q\ell p$ (case drain flowrate) as ΔP is varied in

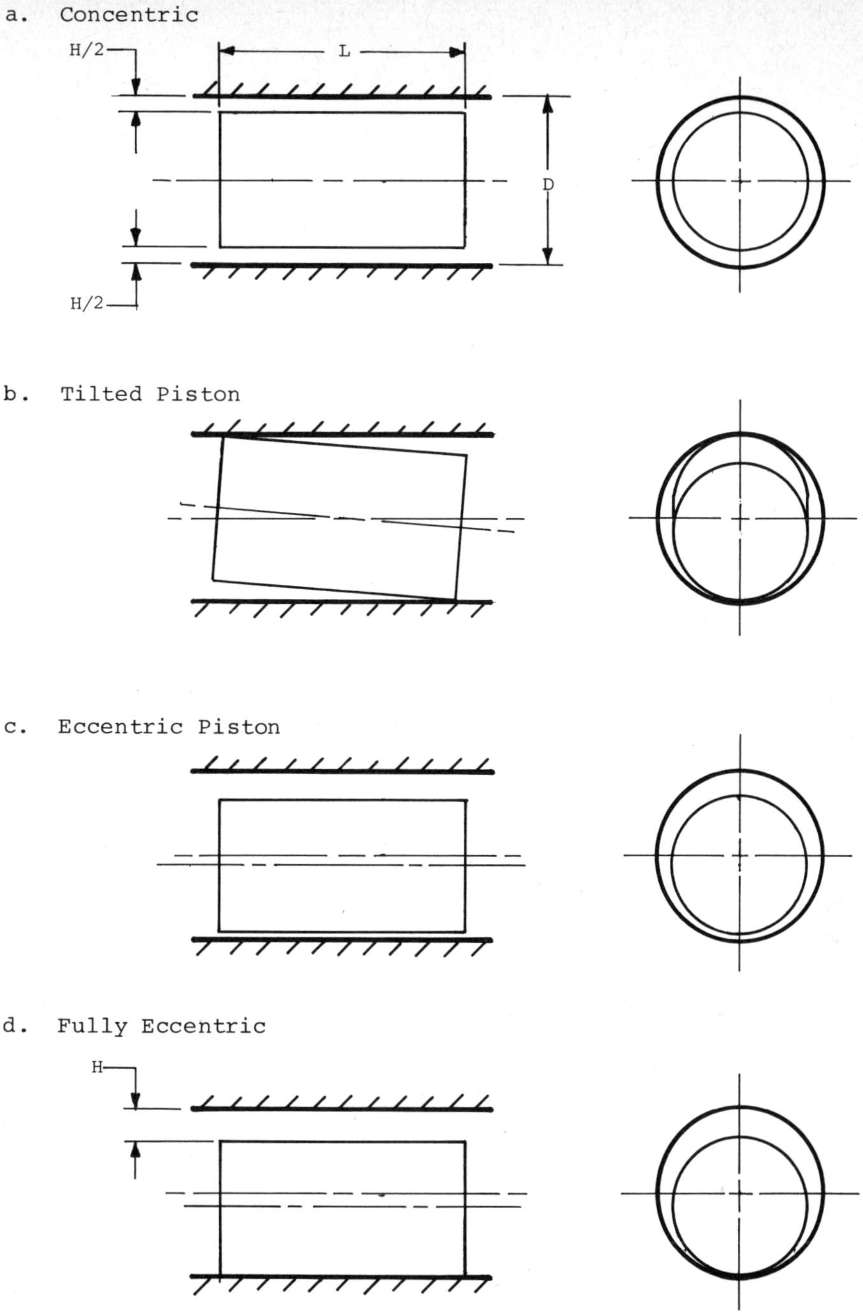

FIG. 7.6 LEAKAGE PASSAGE FOR HYDRAULIC PISTONS

discrete steps. The pump should be running while the test is carried out. Reliable pump manufacturer's data for volumetric efficiency can also be used to estimate $R\ell p$. The same applies to other components. The experienced designer-analyst will not have much difficulty in establishing reasonable $R\ell$ values for use in equation 7.10 and in the related simulation.

For example, as described in Ref. 7.4, the $R\ell p$ value for a 40 kW Rexroth variable displacement axial piston pump was established from data supplied by the manufacturer to be

$$R\ell p = 0.505 \cdot 10^{12} \ \mathrm{Pa/(m^3/s)}$$

7.6 RELIEF VALVE FLOWRATES

7.6.1 Relationships

A pressure relief valve is intended to remain shut up to a certain set pressure, beyond which it will open enough to drain off sufficient flowrate to limit the pressure in the line it is protecting to the set pressure. For use in bond graphs, it is required to have the relationship

$$Qr = f(P) \qquad 7.13(a)$$

where Qr is the flowrate through the relief valve to tank
P is the pressure experienced by the relief valve.

The relief valve itself will have internal dynamics, as it is composed essentially of a damped-spring-mass excited by an external force (pressure • effective area) . Flow forces will also affect the dynamics of relief valve components. However, a well-designed relief valve will have a speed of response many times faster than the response speed capability of the system's driven load [7.1]. That is, the natural frequency of the relief valve will be far higher than that of the driven load system. Bearing in mind that the system designer-analyst's primary concern will be in predicting load response to command inputs, the dynamics of the relief valve can be neglected unless a specific study is being made of the relief valve itself.

Fig. 7.7 illustrates the form of the static $Qr(P)$ relationship to be expected from a relief valve. The relationship can be described by the discontinuous two-part equation

$$\begin{aligned} Qr &= 0 && \text{if } P \leqslant Pset \\ &= 1/Rr \cdot (P - Pset) && \text{if } P > Pset \end{aligned} \qquad 7.13$$

where $1/Rr$ is the slope $\Delta Q/\Delta P$ of the $Qr(P)$ line for $P > Pset$.

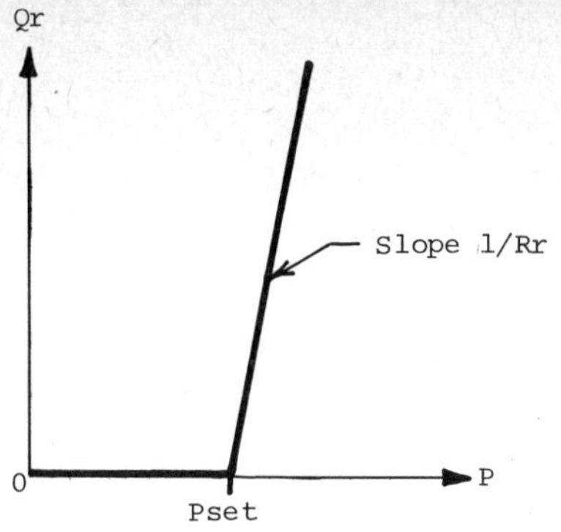

FIG. 7.7 PRESSURE-FLOWRATE CHARACTERISTIC OF A PRESSURE RELIEF VALVE

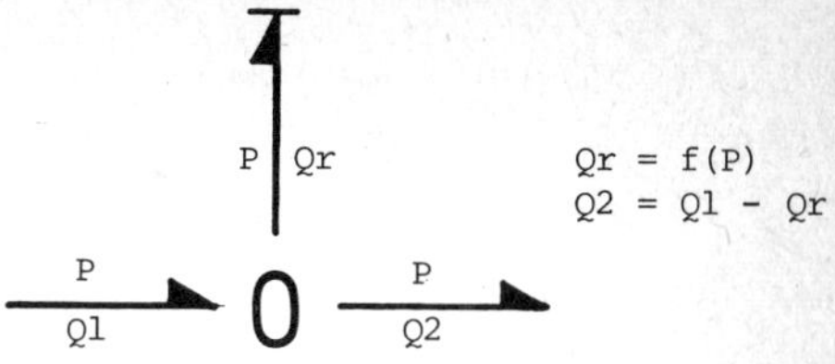

FIG. 7.8 REPRESENTING A PRESSURE RELIEF VALVE BY BOND GRAPH

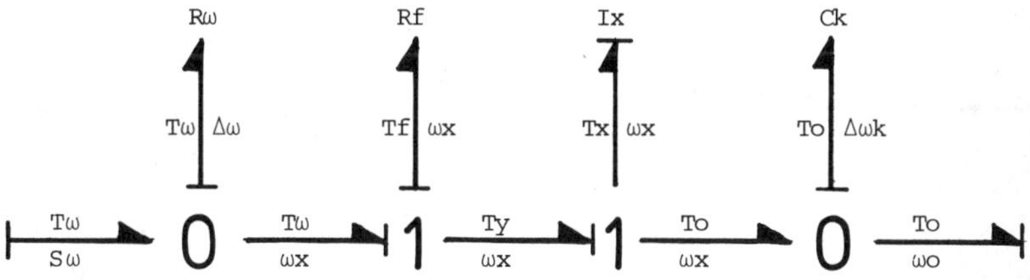

FIG. 7.9 BOND GRAPH FOR AN ELECTRICAL INDUCTION MOTOR (REF. FIG. 5.4)

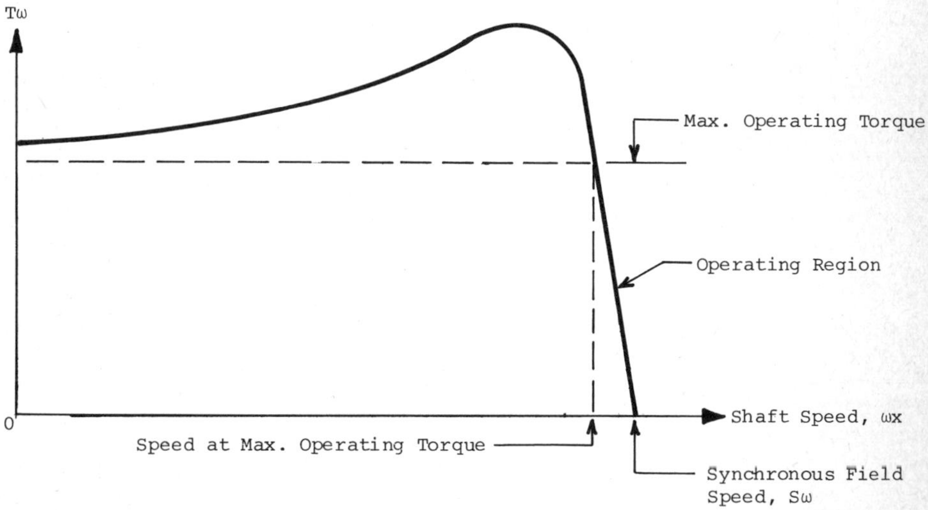

FIG. 7.10 TORQUE-SPEED CHARACTERISTIC FOR AN ELECTRICAL INDUCTION MOTOR

The characteristic of Fig. 7.7 is readily established by experiment. The designer-analyst can expect that a relief valve manufacturer can supply it. The $P > Pset$ characteristic is not always a perfect straight line. It may have a curve to it, in which case the curve can be approximated by a straight line, described by a non-linear relationship, or represented by data points. Alternatively, experimental data for $Qr(P)$ can be used in a simulation. Ref. 7.5 describes in more detail the case for representing relief valves by equation 7.13 in system simulations. It includes the establishing of Rr values for a number of different valves, both direct-acting and pilot-operated. It describes also the results of simulations using the equation 7.13 representation of relief valves, and their support by experiment.

In the occasional event that an analyst may wish to study the dynamics of the relief valve, for example to study chatter, a number of references are available (Ref. 7.6 for example).

In summary, for general system analysis a pressure relief valve can be represented as an R-effect leading from an 0-junction in a bond graph, as illustrated on Fig. 7.8, with the Rr effect to be described by equation 7.13. Note that the single causal bar allowed at the 0-junction can be on either of the main power flow bonds, depending on the particular system the relief valve is used in.

7.6.2 Coefficients

As an example of Rr values, the system of Ref. 7.5 used six different relief valves, all rated in the region 4 to 20 MPa set pressure, and 60 to 180 ℓ/min. Values for Rr ranged from approximately 0.5 to 5×10^9 Pa/(m^3/s) depending upon the particular valve and the particular set pressure. Values for Rr should be available from the relief valve manufacturer, or should be assessable from his valve's performance data.

7.7 ELECTRIC INDUCTION MOTOR

7.7.1 Relationships

Many hydraulic pumps are driven by induction electric motors. Often the motor's output speed can be regarded as constant, in which case the motor's internal characteristics and dynamics can be neglected and the input speed to the pump can be taken as a source (constant). However, by its very nature, an induction motor's output speed varies with load. Sometimes this "speed slip" will be significant to a particular hydraulic system response analysis, and it is then necessary to have a dynamic model of the induction motor. In Chapter 5, Section 5.3.3 and Fig. 5.4, a detailed bond graph was developed for this situation. Fig. 7.9 is a reproduction of Fig. 5.4b with the gyration from source voltage to source motor field speed omitted as redundant. It allows for

- speed slip, as the effect $R\omega$
- friction, as the effect Rf
- motor inertia, as the effect Ix
- output shaft compliance, as the effect Ck .

A typical induction motor slip characteristic is illustrated on Fig. 7.10. Over the operating range the curve can be approximated by

$$T\omega = R\omega \cdot \Delta\omega \tag{7.14}$$

where $T\omega$ is the torque developed on the rotor
$R\omega$ is the slope of the $T\omega - \omega$ relationship
$\Delta\omega$ is the speed slip relative to synchronous field speed $S\omega$.

It is common to allow for friction and windage in a loaded motor as

$$Tf = Kf \cdot \omega x^3 \tag{7.15}$$

where Tf is the torque component of $T\omega$ lost in overcoming friction
Kf is a friction coefficient
ωx is motor shaft speed

The inertia effect of the motor can be allowed for as

$$\omega x = 1/Ix \cdot \int Tx \, dt + \omega x(o) \tag{7.16}$$

where ωx is motor speed
Ix is inertia of rotor, including output shaft
Tx is that torque component of $T\omega$ which is absorbed accelerating inertia Ix .

The shaft compliance effect, Ck , can be described

$$To = 1/Ck \cdot \int \Delta\omega k \, dt + To(o) \tag{7.17}$$

where To is the net torque available at the output end of the shaft
Ck is the shaft compliance (inverse of stiffness)
$\Delta\omega k$ is the angular speed difference, due to deflection, between the motor end and the output end of the shaft.

To and ωo are respectively the torque and the speed available to drive the component to be connected to the motor, and the power available is $To \cdot \omega o$. It should be appreciated that all of the variables in equations 7.14 to 7.17 are dynamic. They could all be expressed as function of time —— i.e. $T\omega(t)$, $\Delta\omega(t)$, $Tf(t)$, etc. —— and should always be viewed as such.

7.7.2 Model

The complete model for the induction motor represented by Fig. 7.9 becomes

Source:	$S\omega$	=	constant
$R\omega$:	$T\omega$	=	$R\omega \cdot \Delta\omega$
Rf:	Tf	=	$Kf \cdot \omega x^3$
Ix:	ωx	=	$1/Ix \cdot \int Tx\, dt + \omega x(o)$
Ck:	To	=	$1/Ck \cdot \int \Delta\omega k\, dt + To(o)$
0-junction:	$\Delta\omega$	=	$S\omega - \omega x$
1-junction:	Ty	=	$T\omega - Tf$
1-junction:	Tx	=	$Ty - To$
0-junction:	$\Delta\omega k$	=	$\omega x - \omega o$

Note that the model is formed on the basis that it will be loaded; i.e. that it will be driving some component, such as a hydraulic pump, which will demand power in the form of torque and speed.

7.7.3 Coefficients

For the model presented, the coefficients required before a simulation can proceed are $S\omega$, $R\omega$, Kf, Ix, and Ck, and the initial values $\omega x(o)$ and To(o). Synchronous field speed $S\omega$ and speed slip coefficient $R\omega$ are basic motor characteristics available with a particular motor. A reputable manufacturer should also be able to supply on request the inertia of the motor's rotor and its output shaft, to which can be added the designer-analyst's calculation of the inertia of any coupling and driven component shafting present in the drive. The friction coefficient Kf can be assessed either from the manufacturer's data or from mechanical efficiency curves for the motor. Shaft compliance Ck can be calculated from conventional torsional stiffness or deflection formula.

As an example, the characteristic coefficients for a particular 50 kW, 240 volt, 50 Hz, 4-pole General Electric induction motor were found to be

$S\omega$ = 1500 rpm

$R\omega$ = 53 $N \cdot m/(rad/s)$

Kf = $0.316 \cdot 10^{-3}\ N \cdot m/(rad/s)^2$

Ix = $16.5 \cdot 10^{-3}\ kg \cdot m^2$

Ck = $10 \cdot 10^{-6}\ rad/(N \cdot m)$

7.7.4 Conclusion

The electric induction motor has been modelled in some detail. The model developed and the particular coefficients for it will be utilized in an example in Chapter 8 where motor dynamics were considered to be significant to the analysis. It is emphasized, again, that it is rare that electric motor dynamics are of primary significance in the dynamic performance analysis of hydraulic control systems.

7.8 HYDRAULIC PUMPS

7.8.1 Relationships

A hydraulic pump normally is running continuously. In dynamic analyses where driven load response is the main objective of study it can often be assumed that pump speed is constant. For fixed displacement pumps, inertia of the pump's moving parts can then be neglected. So too can internal friction in the pump. That is, the pump is regarded as experiencing a speed source, and can be represented by the bond graph of Fig. 7.11a, where

- T is the torque developed at the pump shaft
- Sωp is the pump shaft speed, here regarded as a constant and hence as source
- Vp is the geometric displacement per radian rotation of the pump
- Pp is the pressure at pump discharge
- Qp is the ideal or geometric discharge flowrate of the pump, here regarded as a constant and hence as source SQp
- Rℓp denotes that internal pump leakage is to be considered
- Qℓp is pump leakage flowrate
- Qp is actual pump discharge flowrate.

For a particular drive speed ωp and swept volume rate Vp the bond graph can be simplified to that of Fig. 7.11b.

It is of interest and significance that the causal bar on the pump's output bond is at the left, indicating that pressure is a back effect, to be induced by whatever impedance the pump's discharge flowrate experiences. Thus, the bond graph indicates unambiguously that a positive displacement pump is primarily a generator of flowrate, and not of pressure.

There may be situations in which it is necessary or desirable to include some or all of a fixed displacement pump's own dynamic effects in a simulation. This would become necessary if it was thought that a pump's drive speed could vary significantly under the conditions relevant to a particular performance analysis. Fig. 7.11c shows a bond graph structure which allows for pump friction (Rfp) and pump inertia (Ip).

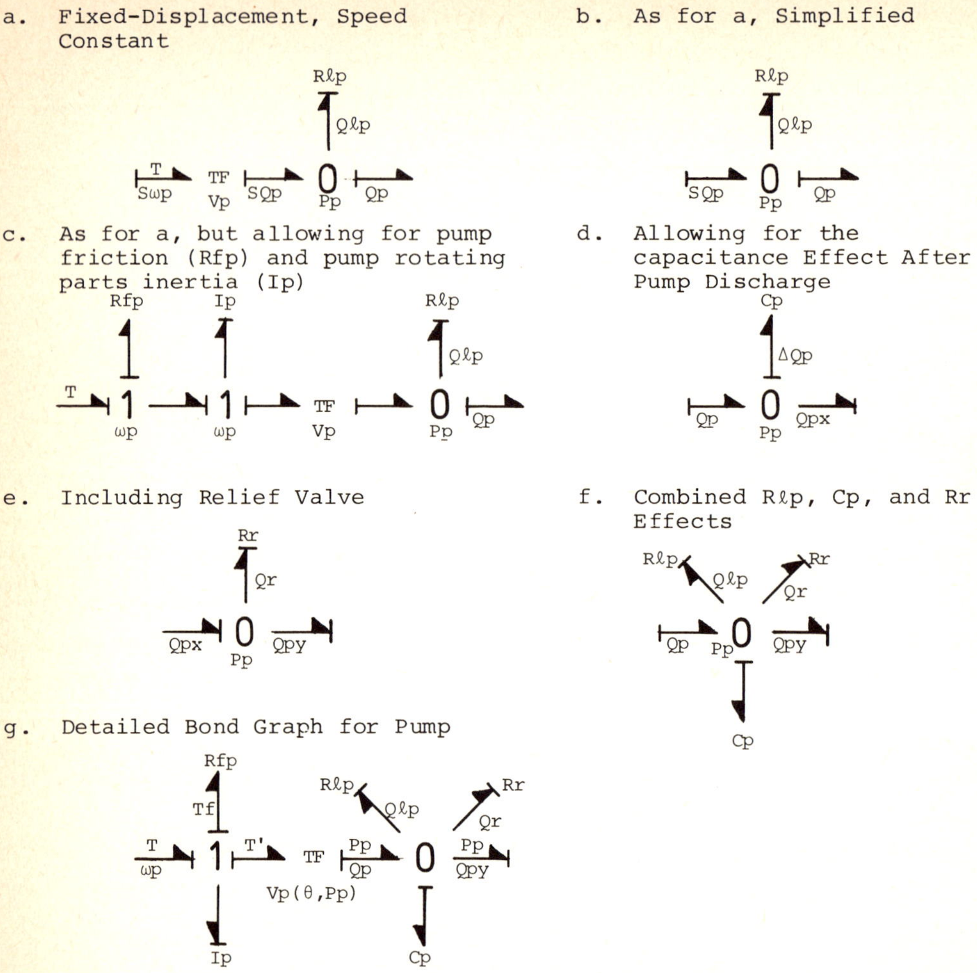

FIG. 7.11 VARIOUS BOND GRAPH CONFIGURATIONS FOR HYDRAULIC SYSTEM PUMPS

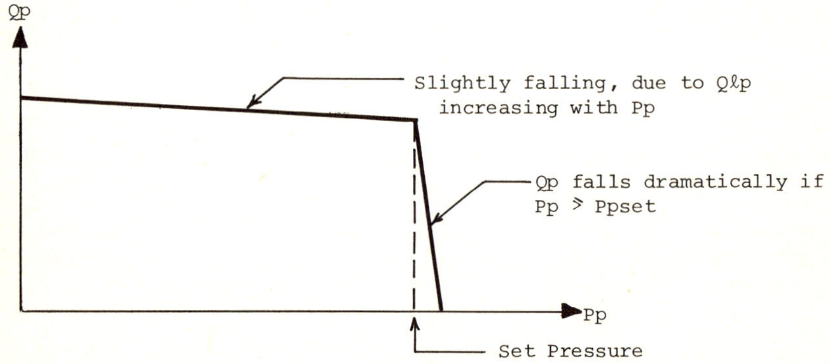

FIG. 7.12 DESIRED CHARACTERISTIC OF PRESSURE-COMPENSATION OF PUMP

Note that input speed can no longer be a source, as ωp has to be allowed to be affected by Rfp and/or Ip . In general neither input torque nor input speed will be a source in such cases. The input bond causality shown on Fig. 7.11c would have to be satisfied by the output bond of the device driving the pump.

For a variable displacement pump, the displacement per radian of the pump is modulated by a suitable mechanism. For a swashplate pump for example Vp on Fig. 7.11a and c would have to be replaced by

$$Vp = Vp' \cdot \theta$$

where Vp' is the volumetric displacement of the pump per radian of pump rotation per radian of swashplate angle measured from zero (swashplate vertical, piston stroke zero)

θ is the swashplate angle measured from zero.

If allowing for variable displacement, the inertia of the swashplate and its driving mechanism may have to be considered, whether or not an assumption of constant pump rotational speed is assumed.

A pump invariably discharges into some kind of a volume. The volume will include the pump's own porting and outlet, and can include the volumes of connecting hydraulic lines, of a high-pressure filter, and an accumulator. The capacitance effect of the pump's discharge side volumes can be recognized as a C effect, as illustrated on Fig. 7.11d, where

Cp recognizes the capacitance effect of the discharge side immediately adjacent to the pump

ΔQp is the amount of Qp 'lost' to the Cp effect

Qpx is the effective volumetric flowrate experienced by the next effect or component to be added to the bond graph.

The majority of hydraulic systems include a pressure relief valve adjacent to the pump, and this can be allowed for as an Rr effect as shown in Fig. 7.11e, where

Rr denotes the presence of the relief valve

Qr is the discharge to tank through the relief valve,

both as discussed in Section 7.6, and

Qpy is the effective flowrate received by the next component or effect to be added to the bond graph.

Either, both, or neither Cp and Rr may exist in a particular analysis.

Remembering that adjacent similar junctions can be combined, leakage, discharge side capacitance, and relief valve can be combined into Fig. 7.11f. Note that if the Cp effect is considered negligible, the causal bar on the output bond Pp • Qpy must revert to the left end of the bond in order to provide the single causal bar needed at the 0-junction.

Fig. 7.11g shows a detailed and general bond graph for a positive displacement pump, with power in as T • ωp and out as Pp • Qpy . The pump's essential power-transforming characteristic is shown as the TF Vp(θ, Pp) , where Vp(θ, Pp) allows that a pump's geometric displacement rate can be a function of θ , a displacement control parameter such as swashplate angle, and also of the pump delivery pressure Pp in the case of pressure compensated pumps.

Including internal leakage, a pressure-compensate pump has a Qp - Pp relationship of the form idealized on Fig. 7.12. At constant pump speed, Qp is insensitive to pressure, in the manner of a normal positive displacement pump, until a set pressure Pset is reached. Pressure in excess of Pset is used to activate a mechanism which causes pump displacement and consequently Qp to decrease very sharply. When Pp falls below Pset , the positive displacement pump characteristic is resumed. The relationship of Fig. 7.12 is readily described as

$$\begin{aligned} Qp &= Vp \cdot \omega p && \text{for } Pp < Pset \\ &= Vp \cdot \omega p + Kp(Pp - Pset) && \text{for } Pp > Pset \end{aligned} \qquad 7.18$$

where Kp is the slope of the Pp > Pset part of Fig. 7.12. Note that Kp is a negative coefficient, of units $(m^3/s)/Pa$.

Most of the R , C , and I effect relationships implied on Fig. 7.11 have already been discussed in the present Chapter. Only pump friction requires further discussion.

Pump friction arises from

- motion between mated mechanical parts within the pump, such as pistons in their bores, cylinder blocks on port plates, piston-to-swashplate bearings, gears running between side plates, vanes on their running tracks, etc.
- shaft bearings
- movement of control mechanisms in the pump
- fluid shear within the pump due to moving parts.

Precise component-by-component assessment of friction in pumps is a complicated process. However, it can be appreciated that the torque required to overcome friction is affected by both pump speed, and pump pressure. The moving parts will be immersed in oil, providing good lubrication. Pump speed is likely to induce a viscous friction effect, i.e. the higher the speed the higher the frictional torque. Pump pressure will increase the normal loading between mated mechanical parts in relative motion, and thus increase the effort needed to overcome friction. That is, in general,

$$Tfp = f(\omega p, Pp) \qquad 7.19(a)$$

Normally, a pump's speed will not vary much, and thus the ωp component of Tfp will be fairly constant. Pump pressure can vary widely and quickly as a system operates. Thus, the Pp component of Tfp can be dynamic. All things considered, if pump friction needs to be included in a particular dynamic analysis, it can be represented

$$Tfp = Tfp\omega + Kfp \cdot Pp \qquad 7.19$$

where Tfp is the total torque required to overcome friction
Tfpω is a constant component of Tfp
Kfp is a coefficient
Pp is pump delivery pressure.

7.8.2 Coefficients

A specific pump's geometric parameters, such as Vp , Ip , and Kp are obtainable from manufacturer's data. Leakage was described in Section 7.5, and capacitance in Section 7.2. Internal friction coefficients Tfpω and Kfp are more difficult to specify precisely, although an assessment of overall friction can be made from mechanical efficiency data.

7.9 4-WAY CONTROL VALVES

7.9.1 Introduction

Because of

- its position as the controlling element of many hydraulic control systems
- its significance to the dynamic response of driven loads
- its numerous design and performance variations

the 4-way control valve will be discussed in some detail.

The 4-way valve has four ports through which power can enter or leave the valve. One port accepts power as an input from the pump system and is designated the supply port. Another port is connected to exhaust (tank). The remaining two, the control ports, typically but not exclusively are connected to the two hydraulic ports of a hydraulic actuator. The direction in which a driven actuator moves is dictated by the direction in which the control valve is actuated. The speed at which a driven actuator moves is dictated by the amount by which the control valve is opened.

There are numerous functional variations of the 4-way control valve. It is usual to characterize a valve by the port interconnections existing when the valve is in its central (neutral) operating state. The most common valve-centre states, illustrated on Fig. 7.13, are

- closed centre: all four ports are closed and are isolated from each other,
- open-centre: all four ports are open to each other,
- tandem-centre: supply port is open to tank (port), but the control ports are both closed.

Various manufacturers have different ways of putting the illustrated porting arrangements into physical hardware. There are centred valve port interconnections which are different from those on Fig. 7.13. A 4-way valve can have two operational positions (no centre position intended), can have three intended positions (valve actuation full left, or central, or full right), or can be intended for continuous actuation (caused to operate in any opening state over its full range of actuation). A 4-way control valve can be of poppet type, or disc type, or spool type. It is not practical to consider all of the variations and ramifications in the present Text. An appreciation of the control relationships for spool-type realizations of the basic configurations of Fig. 7.13 will be provided. From this, development of relationships for variations in the basic valve functions should be practicable.

7.9.2 The Basic Relationship

Flow through a control valve is controlled by the degree of opening of a set of orifices formed by the valve body-spool configuration. The rate of flow through an orifice depends on the area of the orifice and on the pressure drop across it. This functional relationship can be expressed

$$Q = f(A^m, \Delta P^n) \qquad 7.20(a)$$

where Q is flowrate through the orifice
A is the area of orifice opening normal to flow
ΔP is the pressure drop across the orifice
m and n are indices.

In a well-made square-ported spool valve orifice of the type illustrated on Fig. 7.14 it is a reasonable assumption that

$$A = K \cdot X$$

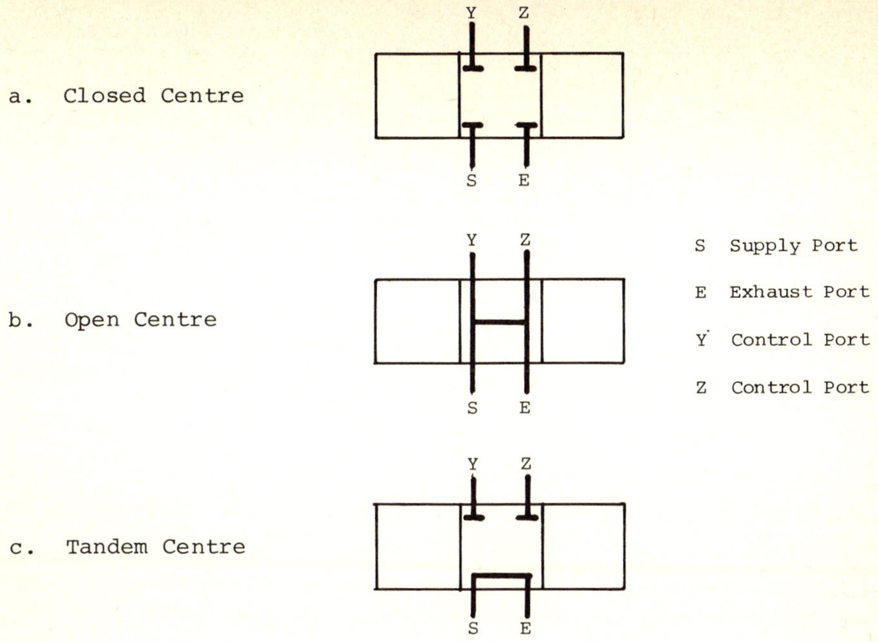

FIG. 7.13 CENTRE ARRANGEMENTS OF COMMON 4-WAY CONTROL VALVES

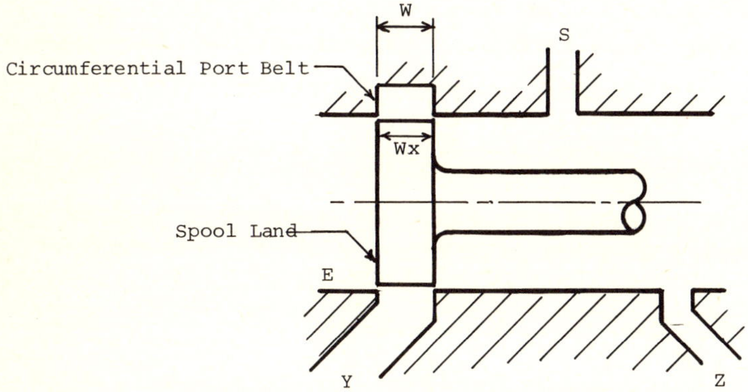

FIG. 7.14 ILLUSTRATING SQUARE-PORTED SPOOL VALVE

- For line-on-line, W = Wx
- For overlap, Wx > W and overlap = 0.5(Wx-W)
- For underlap, Wx < W and underlap = 0.5(W-Wx)
  (Assuming axial symmetry)

$m = 1$

$n = 0.5$ (turbulent flow through an ideal orifice)

giving

$$Q = f(K \cdot X \cdot \Delta P^{\frac{1}{2}}) \qquad 7.20(b)$$

where K is the spool land circumference

X is the axial opening caused by the displacement of the spool in the valve body from the neutral (centre) position.

The functional equation 7.20(b) can usually be expressed

$$Q = Kv \cdot X \cdot \Delta P^{\frac{1}{2}} \qquad 7.20$$

where Kv is an overall coefficient for a particular valve orifice, embracing port circumference and discharge coefficients.

Valve coefficient Kv can vary somewhat with degree of opening X , and with pressure drop ΔP . However, it is often adequate to assume Kv constant at a median value for the particular valve or orifice.

Noting that it expressed flowrate Q as a function of pressure drop ΔP , equation 7.20 can be regarded as a resistance equation, and the valve orifices as R effects. Equation 7.20 is non-linear in two ways; first because $\Delta P^{\frac{1}{2}}$ is not of degree 1 , and second because Q is a function of the product of two variables, X and ΔP . Thus, the "resistance" of the orifice will not be a simple constant as in cases of linear resistance. In essence, the control valve orifices act as variable non-linear resistors which can be defined

$$R = 1/(Kv \cdot X \cdot \Delta P^{-\frac{1}{2}}) \qquad 7.21$$

such that

$$Q = 1/R \cdot \Delta P \qquad 7.22$$

Note that R varies with valve opening X . The larger X , the lower R becomes and the larger the flowrate passed by the orifice, and vice versa.

The idea that each flow orifice in a 4-way control valve acts as a resistor whose resistance is modulated by valve displacement Xv which results in port opening X , is a useful way of understanding the nature of flowrate control imposed by such valves. It is first necessary that all of the flow paths of a particular control valve configuration be recognized. Then the governing flow equations can be formed. This idea will now be developed in turn for the three basic configurations of Fig. 7.13.

7.9.3 Valve Flow Nomenclature

The following nomenclature follows the general nomenclature of Chapter 2. It is detailed here for convenience in appreciating the important characteristics of control valve flows.

Qs	flowrate into the supply port S from outside the valve
Qe	flowrate from the exhaust port E
Qy	flowrate from the control port Y
Qz	flowrate from the control port Z
Qsy	flowrate from supply into the Y port, controlled by orifice Rsy
Qsz	flowrate from supply into the Z port, controlled by orifice Rsz
Qye	flowrate from supply port Y to exhaust controlled by orifice Rye
Qze	flowrate from supply port Z to exhaust controlled by orifice Rze
Qse	flowrate from support port S to exhaust controlled by orifice Rse
Ps	pressure in the supply port S
Pe	pressure in the exhaust port E
Py	pressure in the control port Y
Pz	pressure in the control port Z
Xv	displacement of the spool from its centred position
X	actual axial opening between spool land edge and port
X(o)	amount of spool-to-port underlap, if present
W	axial width of control port circumferential belt in valve body
Wx	axial width of spool land ($\therefore$ overlap = 0.5 (Wx - W))
Kv	flow coefficient for control ports
Kse	flow coefficient for supply-to-exhaust passage, if present
R	denotes a resistive effect

It should be appreciated that except for coefficients X(o) , W , Wx , Kv , Kse , the quantities listed are dynamic, i.e. it is expected that they will vary during dynamic response. Exhaust pressure Pe will normally be constant (i.e. a source), and supply pressure Ps may in some circumstances be assumed constant.

7.9.4 The Closed-Centre Control Valve

Fig. 7.15a illustrates the closed-centre valve requirement. Fig. 7.15b illustrates how the requirement can be realized in spool valve hardware. Ideally, this line-on-line configuration allows no flow into or from the control ports while the spool is centred. In practice, some overlap of spool lands over the port edges is necessary to minimize leakage in neutral. This situation will be considered after the line-on-line case.

Allowing that the spool can be moved either left or right of centre, there are four flow controlling restrictions available in the valve, as illustrated on Fig. 7.15c. If the spool is central, each of the four possible flow paths have infinite resistance, and no flow takes place. If the spool is moved left of centre, Xv positive Fig. 7.15d, the resistance of openings Rsy and Rze becomes finite, allowing flow, while the resistances of openings Rye , Rsz remain infinite, preventing flow. Thus, flow can take place from supply through control port Y , and through control port Z to exhaust. Had the spool been moved right of centre

(Xv negative), Rsz and Rye would become finite, allowing flow from supply to port Z and from port Y to exhaust.

Fig. 7.15e illustrates the flow paths of the 4-way closed-centre valve in a rectangular form which is useful for constructing the valve's bond graph.

Fig. 7.15f shows a bond graph representation of the closed-centre valve function. It is axiomatic that the four control resistances are each a function of spool displacement Xv . The basic equations implied on the bond graph are

Effect	*Equation*	
Rsy; $Qsy = f(\Delta Psy)$:	$Qsy = Kv \cdot Xv \cdot \Delta Psy^{\frac{1}{2}}$	7.23(a)
Rsz; $Qsz = f(\Delta Psz)$:	$Qsz = Kv \cdot Xv \cdot \Delta Psz^{\frac{1}{2}}$	7.23(b)
Rye; $Qye = f(\Delta Pye)$:	$Qye = Kv \cdot Xv \cdot \Delta Pye^{\frac{1}{2}}$	7.23(c)
Rze; $Qze = f(\Delta Pze)$:	$Qze = Kv \cdot Xv \cdot \Delta Pze^{\frac{1}{2}}$	7.23(d)
Supply 0-junction:	$Qs = Qsy + Qsz$	7.23(e)
Exhaust 0-junction:	$Qe = Qye + Qze$	7.23(f)
Y 0-junction:	$Qy = Qsy - Qye$	7.23(g)
Z 0-junction:	$Qz = Qsz - Qze$	7.23(h)
Rsy 1-junction:	$\Delta Psy = Ps - Py$	7.23(i)
Rsz 1-junction:	$\Delta Psz = Ps - Pz$	7.23(j)
Rye 1-junction:	$\Delta Pye = Py - Pe$	7.23(k)
Rze 1-junction:	$\Delta Pze = Pz - Pe$	7.23(ℓ)

The junction equations, (a) to (ℓ), are general. However, the R equations, (a) to (d), are in practice discontinuous. Each needs to be expanded to embrace physical reality, as follows

$$Qsy = \begin{cases} 0 & \text{if Xv is 0} \\ Kv \cdot Xv \cdot \Delta Psy^{\frac{1}{2}} & \text{if Xv is positive and} < W \\ Kv \cdot W \cdot \Delta Psy^{\frac{1}{2}} & \text{if Xv is positive and} > W \\ 0 & \text{if Xv is negative} \end{cases} \qquad 7.23(m)$$

$$Qsz = \begin{cases} 0 & \text{if Xv is 0} \\ 0 & \text{if Xv is positive} \\ Kv \cdot |Xv| \cdot \Delta Psz^{\frac{1}{2}} & \text{if Xv is negative and} < W \\ Kv \cdot W \cdot \Delta Psz^{\frac{1}{2}} & \text{if Xv is negative and} > W \end{cases} \qquad 7.23(n)$$

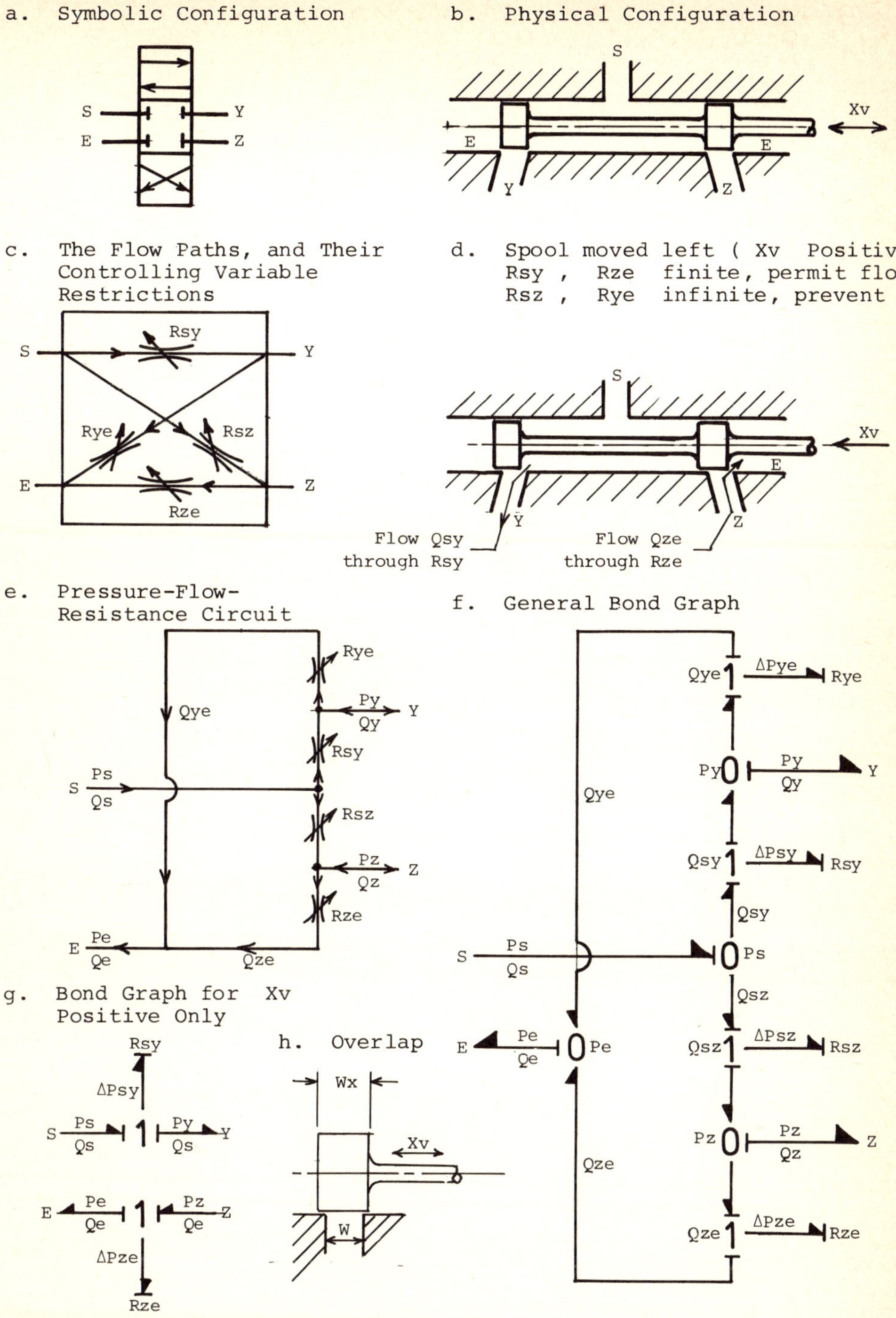

FIG. 7.15 DEVELOPMENT OF BOND GRAPH FOR CLOSED-CENTRE 4-WAY CONTROL VALVE

$$
\left.\begin{aligned}
Qye &= 0 && \text{if } Xv \text{ is } 0 \\
&= 0 && \text{if } Xv \text{ is positive} \\
&= Kv \cdot |Xv| \cdot \Delta Pye^{\frac{1}{2}} && \text{if } Xv \text{ is negative and} \leq W \\
&= Kv \cdot W \cdot \Delta Pye^{\frac{1}{2}} && \text{if } Xv \text{ is negative and} > W
\end{aligned}\right\} \quad 7.23(o)
$$

$$
\left.\begin{aligned}
Qze &= 0 && \text{if } Xv \text{ is } 0 \\
&= Kv \cdot Xv \cdot \Delta Pze^{\frac{1}{2}} && \text{if } Xv \text{ is positive and} \leq W \\
&= Kv \cdot W \cdot \Delta Pze^{\frac{1}{2}} && \text{if } Xv \text{ is positive and} > W \\
&= 0 && \text{if } Xv \text{ is negative}
\end{aligned}\right\} \quad 7.23(p)
$$

Equations 7.23(e) to (p) comprise the model.

If spool-to-port overlap is present, the bond graph is unaffected. However, equations (m), (n), (o), and (p) must be appropriately modified. For example, if W is port width, Wx is land width, and overlap is symmetrical (Fig. 7.15h), then overlap is $0.5\ (Wx - W)$ and the port opening becomes

$$
\begin{aligned}
X &= 0 && \text{for } |Xv| < 0.5\ (Wx - W) \\
X &= Xv - 0.5\ (Wx - W) && \text{for } Xv > 0.5\ (Wx - W) \text{ and } Xv \text{ positive} \\
X &= Xv + 0.5\ (Wx - W) && \text{for } |Xv| > 0.5\ (Wx - W) \text{ and } Xv \text{ negative}
\end{aligned}
$$

and these expressions must be used for Xv in equations 7.23(m), (n), (o), and (p).

If control valve actuation is restricted, as it often is for a particular system response analysis, the general bond graph of Fig. 7.15f can be greatly simplified. For example, for Xv positive only the simple bond graph of Fig. 7.15g is adequate. It implies the necessary equations

$$
\left.\begin{aligned}
Qs &= Qsy = f(\Delta Psy) = Kv \cdot Xv \cdot \Delta Psy^{\frac{1}{2}} \\
Qe &= Qze = f(\Delta Pze) = Kv \cdot Xv \cdot \Delta Pze^{\frac{1}{2}} \\
\Delta Psy &= Ps - Py \\
\Delta Pze &= Pz - Pe
\end{aligned}\right\} \quad 7.24
$$

This is a good demonstration of the advice "use the simplest model which is adequate to the particular task".

7.9.5 The Open-Centre Control Valve

Fig. 7.16a illustrates the centred state of the open-centre valve. Fig. 7.16b illustrates its hardware realization in a spool valve configuration. Spool land underlap relative to the width of the control ports allows all four ports to be open to each other in the centred-spool state. The underlapped spool lands allow the four flowrate control restrictions Rsy, Rsz, Rye, and Rze, all to be finite in the centre state. Normally, though not necessarily, all four R's will be equal in resistance when the spool is central. Normally also, in the central position $Xv = 0$

$$Qs = Qsy + Qsz, \quad \text{and} \quad Qsy = Qsz = Qs/2$$

$$Qye = Qsy, \quad \text{and} \quad Qze = Qsz$$

$$Qe = Qye + Qze, \quad Qe = Qs, \quad \text{and} \quad Qye = Qze$$

$$Qy = Qz = 0$$

If the spool is moved left, the Rsy and Rze openings will increase in area and decrease in resistance to flow, to allow increased flowrate through themselves. However, Rsz and Rye will increase in resistance restricting flow through themselves. Then, Qy and Qz can become finite, Qy positive and Qz negative, and the valve can drive an actuator.

If the spool is moved further left, Rsz and Rye will close, their resistances becoming infinitely high, and flow through them being blocked. Meanwhile, Rsy and Rze can continue to open until their ports are fully open to offer minimum resistance to flow. The valve behaves like a closed-centre valve after Rsz and Rye close.

If the spool is moved right from centre, similar but reversed actions develop.

Note that Qsy describes a flowrate through Rsy from supply into port Y, but is not the flowrate from port Y to outside, as part of Qsy flows through Rye to become Qye. Similarly, Qz and not Qsz is the external flowrate from port Z.

Fig. 7.16c shows a functional layout between the four ports and the four flow-controlling restrictions of the open-centre valve. It is a very useful bridge to formation of the bond graph of Fig. 7.16d. Figs. 7.16c and d are identical with their counterparts in Fig. 7.15 for the closed-centre valve.

General equations for the open-centre valve are, as for the closed-centre valve, discontinuous. For a symmetrically underlapped valve the model comprises the junction equations, identical with those for the closed-centre valve

$$Qs = Qsy + Qsz \tag{7.25(a)}$$

$$Qe = Qye + Qze \tag{7.25(b)}$$

$$Qy = Qsy - Qye \tag{7.25(c)}$$

$$Qz = Qsz - Qze \tag{7.25(d)}$$

$$\Delta Psy = Ps - Py \tag{7.25(e)}$$

$$\Delta Psz = Ps - Pz \tag{7.25(f)}$$

$$\Delta Pye = Py - Pe \tag{7.25(g)}$$

$$\Delta Pze = Pz - Pe \tag{7.25(h)}$$

plus the R effect equations appropriate to open-centre with uniform underlap X(o) and port width W

$$
\left.\begin{array}{lll}
\text{spool centred:} & Qsy = Kv \cdot X(o) \cdot \Delta Psy^{\frac{1}{2}} & \text{if } Xv \text{ is } 0 \\
\text{port part open, positive } Xv: & = Kv \cdot \left(X(o)+Xv\right) \cdot \Delta Psy^{\frac{1}{2}} & \text{if } Xv \text{ is positive but} < \left(W-X(o)\right) \\
\text{port full open:} & = Kv \cdot W \cdot \Delta Psy^{\frac{1}{2}} & \text{if } Xv \text{ is positive but} > \left(W-X(o)\right) \\
\text{port part open, negative } Xv: & = Kv \cdot \left(X(o)-|Xv|\right) \cdot \Delta Psy^{\frac{1}{2}} & \text{if } Xv \text{ is negative but } |{<}|\ X(o) \\
\text{port closed:} & = 0 & \text{if } Xv \text{ is negative but } |{>}|\ X(o)
\end{array}\right] \quad 7.25(i)
$$

$$
\left.\begin{array}{lll}
\text{spool centred:} & Qsz = Kv \cdot X(o) \cdot \Delta Psz^{\frac{1}{2}} & \text{if } Xv \text{ is } 0 \\
\text{port part open, positive } Xv: & = Kv \cdot \left(X(o)-Xv\right) \cdot \Delta Psz^{\frac{1}{2}} & \text{if } Xv \text{ is positive but} < X(o) \\
\text{port closed:} & = 0 & \text{if } Xv \text{ is positive but} > X(o) \\
\text{port part open, negative } Xv: & = Kv \cdot \left(X(o)+|Xv|\right) \cdot \Delta Psz^{\frac{1}{2}} & \text{if } Xv \text{ is negative but } |{<}|\ \left(W-X(o)\right) \\
\text{port full open:} & = Kv \cdot W \cdot \Delta Psz^{\frac{1}{2}} & \text{if } Xv \text{ is negative but } |{>}|\ \left(W-X(o)\right)
\end{array}\right] \quad 7.25(j)
$$

$$
\left.\begin{array}{lll}
\text{spool centred:} & Qye = Kv \cdot X(o) \cdot \Delta Pye^{\frac{1}{2}} & \text{if } Xv \text{ is } 0 \\
\text{port part open, positive } Xv: & = Kv \cdot \left(X(o)-Xv\right) \cdot \Delta Pye^{\frac{1}{2}} & \text{if } Xv \text{ is positive but} < X(o) \\
\text{port closed:} & = 0 & \text{if } Xv \text{ is positive but} > X(o) \\
\text{port part open, negative } Xv: & = Kv \cdot \left(X(o)+|Xv|\right) \cdot \Delta Pye^{\frac{1}{2}} & \text{if } Xv \text{ is negative but } |{<}|\ \left(W-X(o)\right) \\
\text{port full open:} & = Kv \cdot W \cdot \Delta Pye^{\frac{1}{2}} & \text{if } Xv \text{ is negative but } |{>}|\ \left(W-X(o)\right)
\end{array}\right] \quad 7.25(k)
$$

$$
\left.\begin{array}{lll}
\text{spool centred:} & Qze = Kv \cdot X(o) \cdot \Delta Pze^{\frac{1}{2}} & \text{if } Xv \text{ is } 0 \\
\text{port part open, positive } Xv: & = Kv \cdot \left(X(o)+Xv\right) \cdot \Delta Pze^{\frac{1}{2}} & \text{if } Xv \text{ is positive but} < \left(W-X(o)\right) \\
\text{port full open:} & = Kv \cdot W \cdot \Delta Pze^{\frac{1}{2}} & \text{if } Xv \text{ is positive but} > \left(W-X(o)\right) \\
\text{port part open, negative } Xv: & = Kv \cdot \left(X(o)-|Xv|\right) \cdot \Delta Pze^{\frac{1}{2}} & \text{if } Xv \text{ is negative but } |{<}|\ X(o) \\
\text{port closed:} & = 0 & \text{if } Xv \text{ is negative but } |{>}|\ X(o)
\end{array}\right] \quad 7.25(\ell)
$$

Equations 7.25(a) to (j) provide a model allowing for all degrees of valve opening, positive and negative and with port opening saturation. A particular analysis might not require comprehensive description of the valve characteristics, and the bond graph and the model could be simplified accordingly, as indicated with the closed-centre valve.

Particularly with open-centre valves, Kv may not be a single constant. They often have a high gain adjacent to neutral and a lower gain as the spool moves well clear of neutral. Gain here means the $\Delta Q - \Delta Xv$ characteristic for particular values of ΔP .

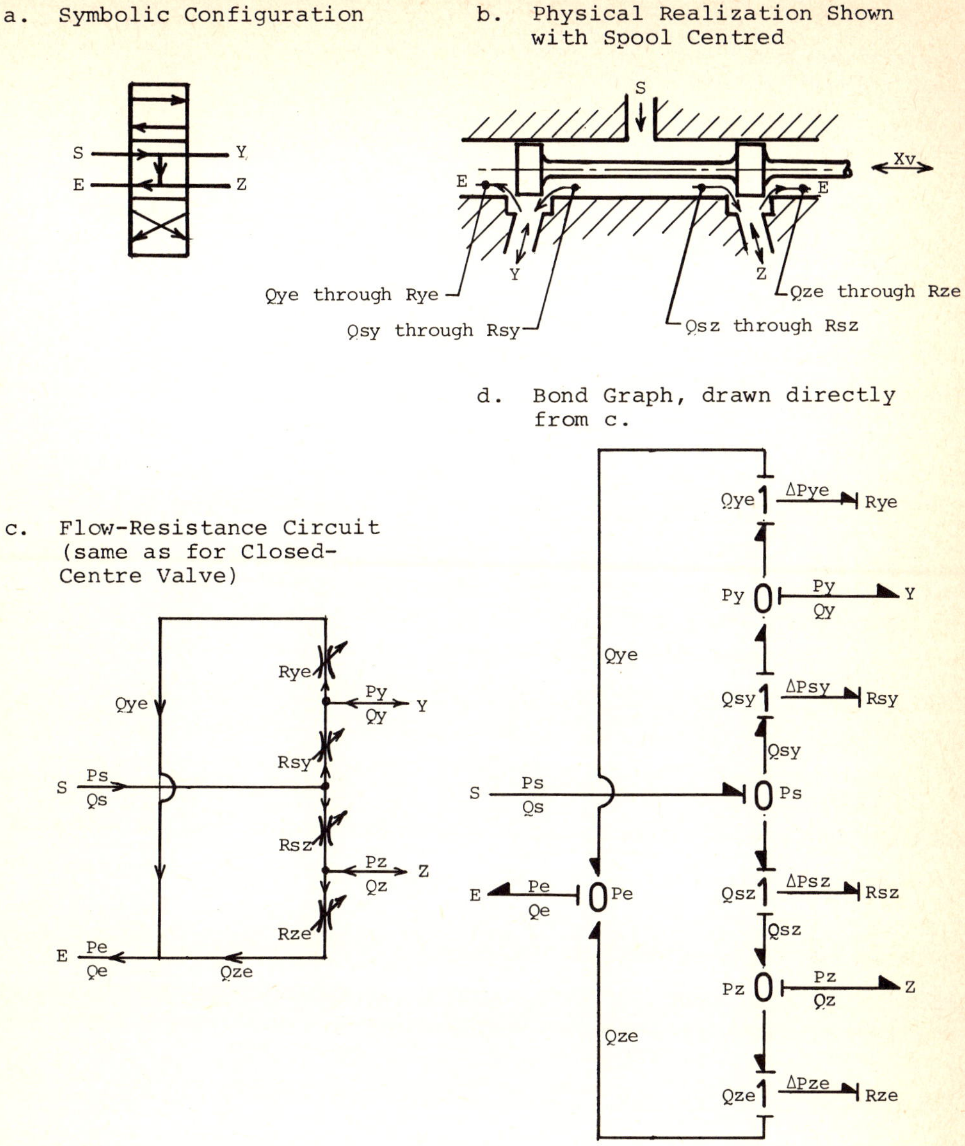

FIG. 7.16 DEVELOPING A BOND GRAPH FOR AN OPEN-CENTRE 4-WAY CONTROL VALVE

7.9.6 The Tandem-Centre Control Valve

Fig. 7.17a illustrates the tandem-centre state, and Fig. 7.17b a realization in spool valve hardware. In the central position the two control ports Y and Z are blocked and isolated while supply can flow fairly freely to exhaust.

The tandem-centre valve has the four flow controlling orifices Rsy , Rsz , Rye , Rze , of the closed and open-centre valves, plus a fifth orifice Rse through which flow can pass from supply to exhaust.

Fig. 7.17c illustrates the orifices and flow paths of a tandem-centre valve, in which

- in the centre position, only Rse is finite, the remaining four restrictions having infinite resistance to flow. Thus only Qse is finite, and Qse = Qs = Qe ,
- as the spool is moved left Rse remains finite though of increasing resistance to flow; Rsy and Rze become finite, permitting flows Qsy and Qze ; Rsz and Rye remain of infinite resistance to flow,
- as the spool is moved further left, Rse becomes of infinite resistance preventing flow Qse ; Rsy and Rze remain finite, and Rsz and Rye remain of infinite resistance,
- the comparable situation for spool displacement to the right of centre can be described.

Fig. 7.17d shows a bond graph for the tandem-centre valve. It is similar to those for the closed and open-centre control valves except that the supply to exhaust path through the centre position is included as R effect Rse . The equations describing the flow possibilities, prepared using the bond graph, are

0-junctions:	Qs	=	Qsy + Qsz + Qse		7.26(a)
	Qy	=	Qsy - Qye		7.26(b)
	Qz	=	Qsz - Qze		7.26(c)
	Qe	=	Qye + Qze + Qse		7.26(d)
1-junctions:	ΔPsy	=	Ps - Py		7.26(e)
	ΔPsz	=	Ps - Pz		7.26(f)
	ΔPye	=	Py - Pe		7.26(g)
	ΔPze	=	Pz - Pe		7.26(h)
	ΔPse	=	Ps - Pe		7.26(i)

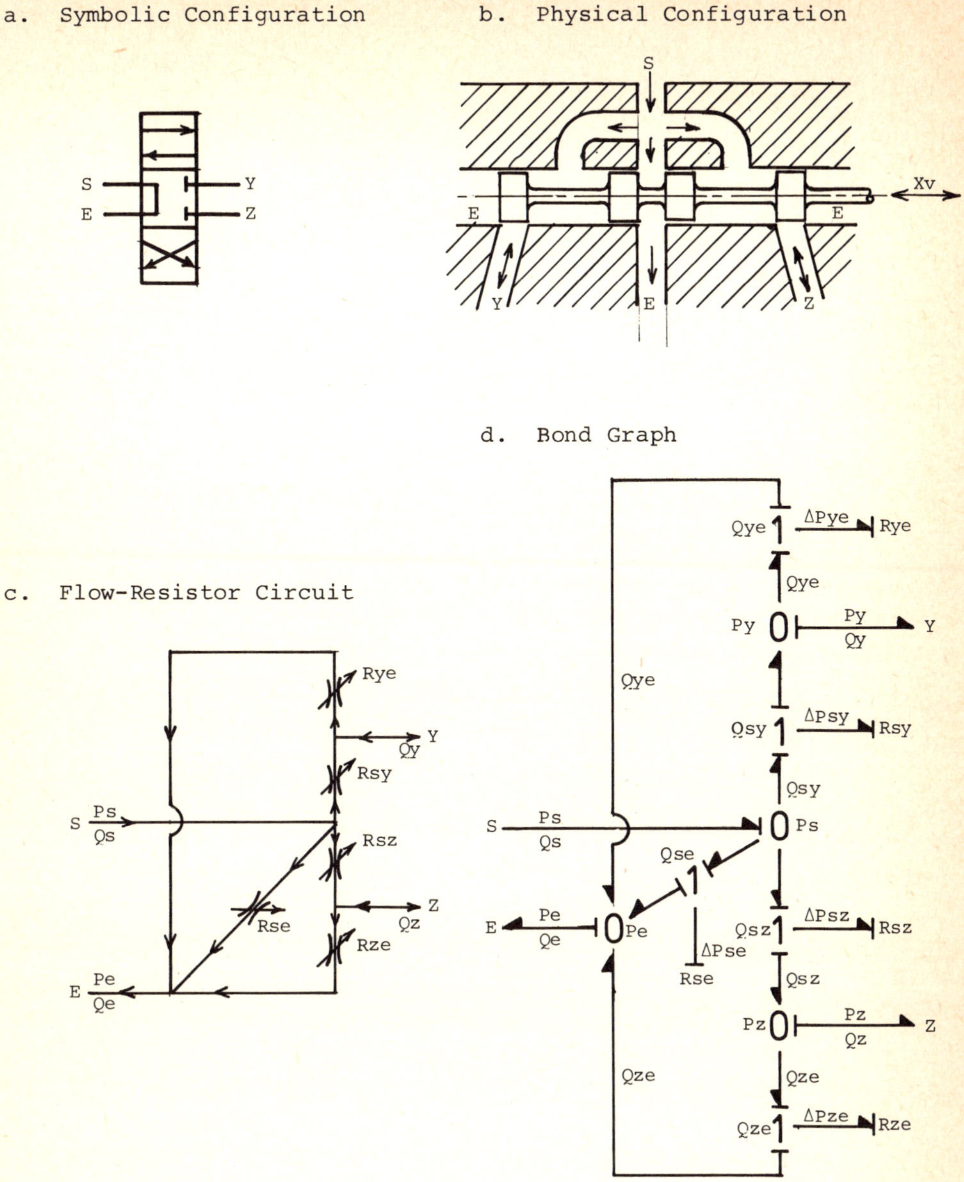

FIG. 7.17 DEVELOPING A BOND GRAPH FOR A TANDEM-CENTRE 4-WAY CONTROL VALVE

R effects:

$$
\left.\begin{aligned}
Qsy &= 0 && \text{if } Xv \text{ is } 0\\
&= Kv\cdot Xv\cdot \Delta Psy^{\frac{1}{2}} && \text{if } Xv \text{ is positive but} < W\\
&= Kv\cdot W\cdot \Delta Psy^{\frac{1}{2}} && \text{if } Xv \text{ is positive but} \geqslant W\\
&= 0 && \text{if } Xv \text{ is negative}
\end{aligned}\right] \quad 7.26(j)
$$

$$
\left.\begin{aligned}
Qsz &= 0 && \text{if } Xv \text{ is } 0\\
&= 0 && \text{if } Xv \text{ is positive}\\
&= Kv\cdot |Xv|\cdot \Delta Psz^{\frac{1}{2}} && \text{if } Xv \text{ is negative but} < W\\
&= Kv\cdot W\cdot \Delta Psz^{\frac{1}{2}} && \text{if } Xv \text{ is negative but} \geqslant W
\end{aligned}\right] \quad 7.26(k)
$$

$$
\left.\begin{aligned}
Qye &= 0 && \text{if } Xv \text{ is } 0\\
&= 0 && \text{if } Xv \text{ is positive}\\
&= Kv\cdot |Xv|\cdot \Delta Pye^{\frac{1}{2}} && \text{if } Xv \text{ is negative but} < W\\
&= Kv\cdot W\cdot \Delta Pye^{\frac{1}{2}} && \text{if } Xv \text{ is negative but} \geqslant W
\end{aligned}\right] \quad 7.26(\ell)
$$

$$
\left.\begin{aligned}
Qze &= 0 && \text{if } Xv \text{ is } 0\\
&= Kv\cdot Xv\cdot \Delta Pze^{\frac{1}{2}} && \text{if } Xv \text{ is positive but} < W\\
&= Kv\cdot W\cdot \Delta Pze^{\frac{1}{2}} && \text{if } Xv \text{ is positive but} \geqslant W\\
&= 0 && \text{if } Xv \text{ is negative}
\end{aligned}\right] \quad 7.26(m)
$$

$$
\left.\begin{aligned}
Qse &= Kse\cdot Wse\cdot \Delta Pse^{\frac{1}{2}} && \text{if } Xv \text{ is } 0\\
&= Kse\cdot (Wse-|Xv|)\cdot \Delta Pse^{\frac{1}{2}} && \text{if } |Xv| \text{ is finite but} < Wse\\
&= 0 && \text{if } |Xv| \text{ is finite but} \geqslant Wse
\end{aligned}\right] \quad 7.26(n)
$$

where line-on-line spool to port configurations have been assumed, Xv is spool displacement from centre, W is port width, and Wse describes the opening of the supply to exhaust port in neutral. The likely difference of the Rse port is indicated by using specific Kse and Wse for it. Like the closed-centre valve, overlap is likely in which case the X term and conditions for equations 7.26(j) to (n) must be modified appropriately. If considered appropriate, the Kv values can be different for each or some of the S , E , Y , and Z ports. Like the previous control valves, the bond graph and its equations are simplified if valve motion is restricted to one side only.

7.9.7 Summary

Bond graphs for closed-centre, open-centre, and tandem-centre 4-way control valves have been presented. From them, equations appropriate to a particular analysis can be written. The forms of the various equations to be expected have been given. Analysis of a specific valve under specific operating conditions must include a careful selection of the bond graph to be used, and equally careful specification of the equation set. The equations must recognize any valve lap present, limits of port width opening, and any unsymmetrical aspects of the specific valve. A restricted analysis often allows considerable simplification of a bond graph and consequently of the equations needed.

Consideration of the bond graphs of Figs. 7.15, 7.16, and 7.17 reveals that Fig. 7.17d could be regarded as a general bond graph for all three centre configurations considered. For the closed and open-centre arrangements, it would be necessary only to specify that the resistive effect Rse is infinitely high for all spool displacement Xv , allowing no flow Qse . The equation set derived from the general bond graph would be different for each of the three cases only in the $Q = f(\Delta P)$ equations.

Fig. 7.17d can be put into the circular form of Fig. 7.18a, or in the form of Fig. 7.18b. This latter form might be the most handy, and will be used in future examples.

Equations 7.20 to 7.26 show the types of equations and equation sets required to describe the valve flows. Specification of valve opening X for each flow equation, in terms of Xv (magnitude and sign), and of particular valve port and lap conditions, is the major challenge. All flows, if finite, are unidirectional except for Qy and Qz , the external flows at the control ports. If Qy is positive, Qz must be negative, and vice versa, and the relevant equations must reflect this.

7.10 ACTUATORS

7.10.1 Introduction

Bond graph structures for a linear actuator were developed in some detail in Chapter 5, Section 5.3.2. Fig. 5.3 showed the developments for the conditions

- back pressure effects neglected
- no leakage across the actuator piston.

Inertia of the actuator's moving assembly (Ia) and its friction (Rfa) were included in Fig. 5.3g, and neglected in Fig. 5.3h. In both cases the capacitance effect of oil flowing into the cylinder is included. A linear actuator is included in the system whose bond graph and equations were developed in Section 5.5.4, with the same limitations associated with Fig. 5.3.

In the present Section, bond graphs for both linear and rotary actuators will be further developed.

7.10.2 Linear Actuator

The back pressure developed in the discharge size of an actuator as its piston is driven in one direction generates a force which opposes motion. This return-side

a. A Circular Form

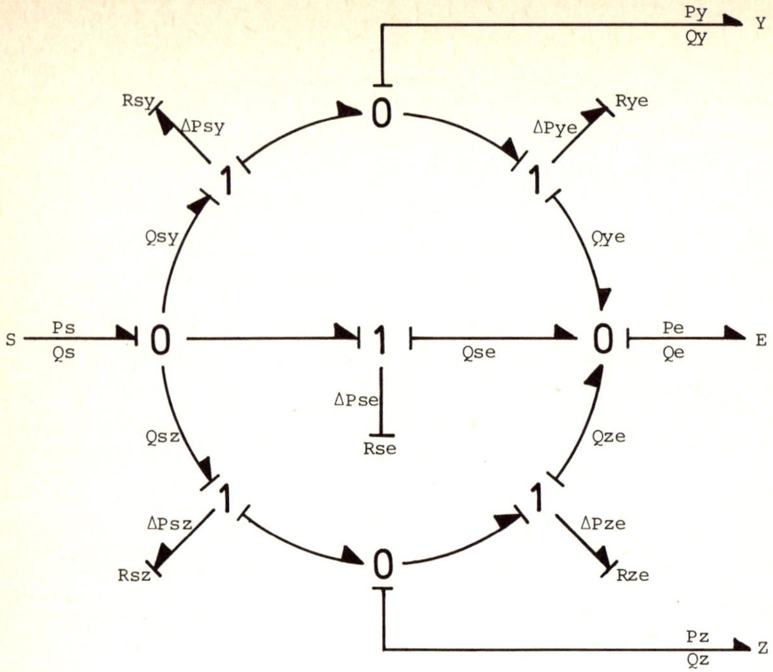

b. A Parallel Form

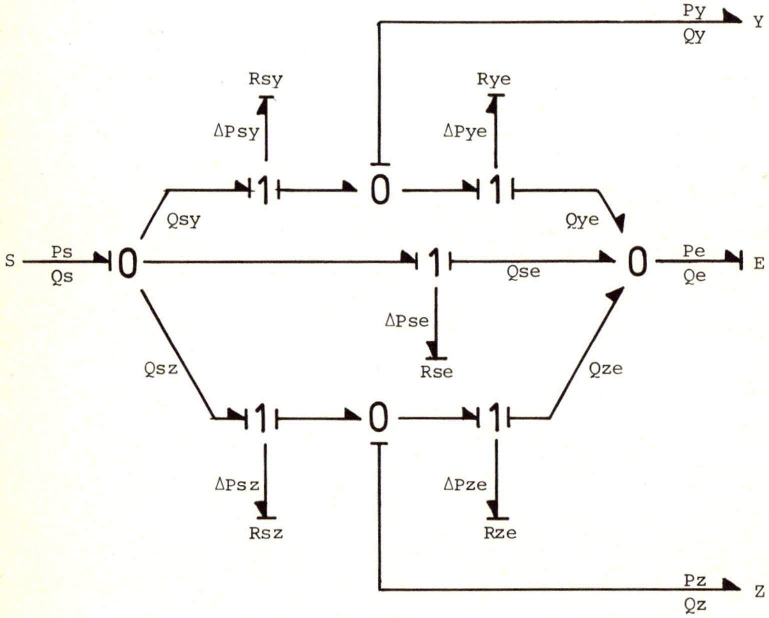

FIG. 7.18 GENERAL BOND GRAPH, 4-WAY CONTROL VALVE

For closed and open centre Rse offers infinite resistance, rendering QSE = 0 for all Xv . Functionally identical with Fig. 7.17d.

effect can be included in a bond graph via the 1-junction shown on Fig. 7.19a, which sums algebraically the forward-side pressure-generated force Fay and the return-side pressure force Faz . The 1-junction merely describes the net pressure force Fa as

$$Fa = Fay - Faz \qquad 7.27$$

Note that the back pressure effect causes a power loss through a loss of effective force.

Fig. 7.19b shows the development of Fig. 5.3g to allow for the back pressure effect. Pressure Paz is generated because flow is being forced from the discharge side of the actuator into the return line by forward motion of the actuator piston. This is clearly a capacitive effect (flow into a C causes effort build-up in the C) and this is recognized by including Caz in the return line bond graph. It is also the reason why back pressure Paz is causal at the TF end of its bond.

If it was decided that actuator friction (Rfa) and inertia (Ia) could be neglected in a particular analysis, or combined with the friction and inertia of a driven load, they and their 1-junctions are simply deleted from Fig. 7.19b.

To allow for the effects of internal leakage across the actuator piston, the bond graph can be modified to that illustrated on Fig. 7.19c. A bond from the 0-junction of Cay , or from a separate 0-junction downstream of it, allows that some of the incoming flow Qay is lost as leakage $Q\ell a$. A bond into the 0-junction of Caz , or into a separate 0-junction upstream of it, allows $Q\ell a$ to be added to Qaz . The 1-junction placed in the $Q\ell a$ flow bonds recognizes that it is pressure drop (Pay - Paz) across the piston that induces leakage.

Fig. 7.19c is a quite detailed model structure for a linear actuator, and is suitable for study of the dynamic performance of the actuator itself. It is suitable also for connection to the bond graphs of inertia loads having compliance between actuator and inertia such as were discussed with Fig. 7.5, Entries 4 to 7. Entry and/or exit pressure loss can be included by adding 1-junctions at the left.

It is important to be aware that Fig. 7.19c and its causalities are not unique. The actuator bond graph can depend also on the type of load it is to drive. For example if it is to drive a simple inertia without compliance in the connection, the causality of the force output bond on Fig. 7.19c will be incorrect. The inertia load, Fig. 7.5 Entry 1, requires the causal bar at the right end of the Fa bond of Fig. 7.19c. A little thought resolves this conflict. If there is no compliance between Ia and Im they should be lumped, reflecting the rigidity of connection. The Ia bond of Fig. 7.19c becomes the actuator's output bond and the former Fa bond is deleted as irrelevant, Fig. 7.19d. The same arrangement applies for the same reason to the load systems represented on Fig. 7.5, Entries 2, 3, and 8. Actuator friction and load friction also can be lumped together if there is no compliance between them, i.e. if their bonds share the same velocity.

Note that the effects Cay and Caz cannot be represented by simple coefficients, except for small perturbation response analysis. In general the cylinder end volumes vary with their displacement as discussed in Section 7.2, and the formulation of the Ca effects should reflect this.

In Chapter 8 there are further examples of various actuator-load situations.

7.10.3 Rotary Actuator (Hydraulic Motor)

The hydraulic motor can be viewed as the rotational equivalent of the linear actuator. Its moving parts have inertia and friction, its chambers capacitance, and it has internal leakage. Fig. 7.19 can be quickly modified to its rotational equivalent by replacing force by torque, velocity by angular velocity, and the TF modulator coefficients A by Va, where Va is the swept volume per radian of rotation of the motor. The TF equations become

$$\text{Pay} \cdot \text{Va} = T \qquad 7.28$$

$$\text{Qay} \cdot \text{Va}^{-1} = \omega \qquad 7.29$$

7.10.4 Equations and Coefficients

The equation forms associated with Fig. 7.19 and its rotational equivalent have all been discussed and will not be re-presented here. Attention is drawn to the variation of the end volumes as the piston of a linear actuator moves. This affects the capacitances of the linear actuator, as discussed in Section 7.2. The varying capacitance effect does not occur to anywhere near the same extent with hydraulic motors and can usually be ignored.

There is usually little difficulty in arriving at the values of coefficients for actuator equations. Most are apparent from manufacturers' data sheets. Leakage coefficients can be inferred from volumetric efficiency data, friction from mechanical efficiency, while geometric data such as piston areas for cylinders, swept volume per radian for motors, and inertias of moving assemblies are supplied directly. This leaves estimation of capacitances Cay and Caz as the main difficulty. This in turn requires estimations of volumes under pressure and effective bulk modulus, both at the specific location being considered, as discussed in Section 7.2.

Loaded hydraulic actuators usually have low damping, with damping ratios of the order of 0.01 to 0.2 depending on the conditions of bearings, seals, and glands.

a. Allowing for Back Pressure Effect

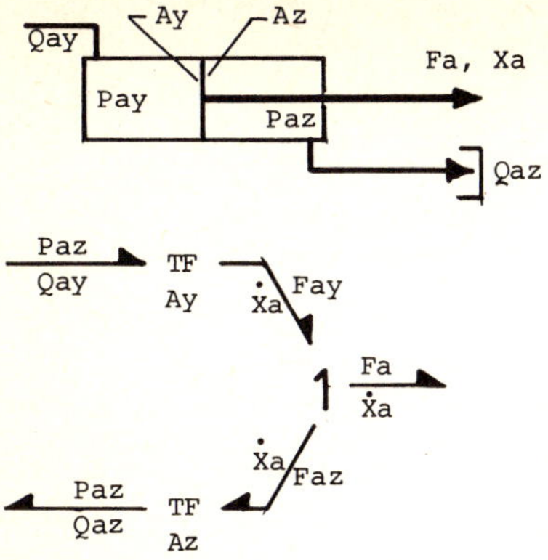

b. Including Actuator Capacitances (Cay and Caz), Friction (Rfa) and Inertia (Ia)

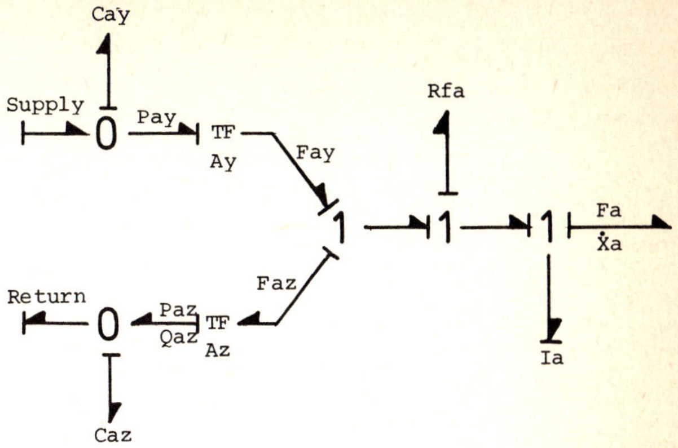

c. Including Internal Cross-Piston Leakage (Qℓa through Rℓa)

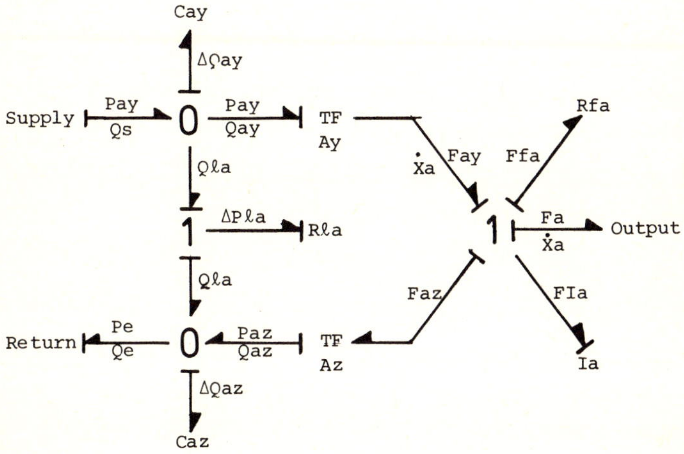

d. Output End of Bond Graph if Actuator Driving a Rigidly Connected Inertia Load Im

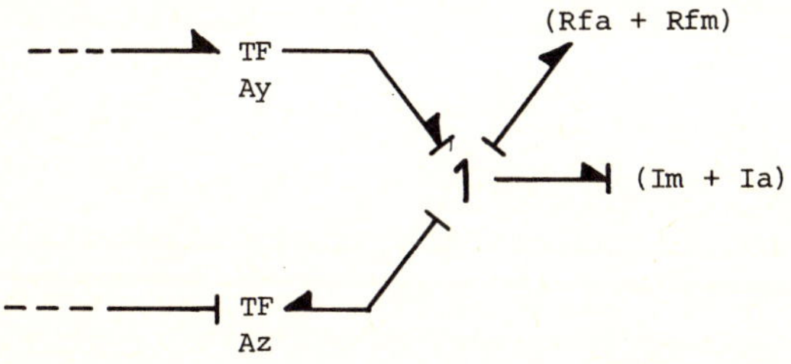

FIG. 7.19 BOND GRAPHS FOR LINEAR ACTUATOR

7.11 SOME OTHER COMMON COMPONENTS (Lines, Filters, Accumulators, Valves)

7.11.1 Hydraulic Lines

In most hydraulic control systems the lines are short and are of reasonably large diameter to keep fluid flow velocities to a low value. In these situations it is usually reasonable to ignore line resistance as negligible relative to, for example, control valve resistance, line capacitance as negligible relative to, for example, actuator capacitance, and fluid inertiance as a very minor dynamic effect relative to the inertia of the actuator and load. Alternatively, line resistance can be lumped with valve resistance and possibly with actuator inlet resistance, and line capacitance with actuator or other adjacent components' capacitance(s). There may be cases however where it is desirable to include some or all of a line's R , C , and I effects in a dynamic model. The situation could arise

- if the lines are long
- if the lines are of small bore
- if the system is being driven in a highly dynamic mode.

Pressure drop in a line often can be accounted for simply by placing a 1-junction in the appropriate power bond. For example, Fig. 7.20a shows line resistance as Rh included in between a control valve (Rv) and an actuator (Ca) . In this particular situation, the Rv and Rh effects come from adjacent 1-junctions and can be shown off the one 1-junction. Rv and Rh really should not be lumped as they are of different causality. For Rv we want $Qv = f(Rv)$ to give the controlled flowrate through the valve. For Rh we want $\Delta Ph = f(Qv)$ to allow for pressure drop in the line. The Rh effect is usually small and if neglected or lumped with Rv no great error in system response results.

Similarly, line capacitance can be represented by Ch leading from an 0-junction in the appropriate power bond, Fig. 7.20b. As with Rh , just where Ch fits in a bond graph depends on the adjacent components. C effects from adjacent 0-junctions (no 1-junctions between) can be lumped. In fact, to preserve integral causality they must be lumped (Fig. 7.20c), reflecting that only one value of the potential variable can exist around an 0-junction C effect. It is proper to separate line capacitance from another capacitance only if there is an R or I effect in between. For example if there is a significant actuator entry pressure drop, Fig. 7.20d could be appropriate. It is of interest to see that entry resistance Ra has the same causality as Rv , requiring description as $Qa = f(\Delta Pa)$ rather than the $\Delta Ph = f(Qv)$ of the line resistance Rh . In essence, Ch becomes the driving component as far as flow into Ca is concerned, and Ra the controlling resistance.

In extreme cases (long small bore lines, highly dynamic flows) the line dynamics can be included by semi-distributed parameter modelling in which the R and C

a. Including Hydraulic Line Resistance (Rh)

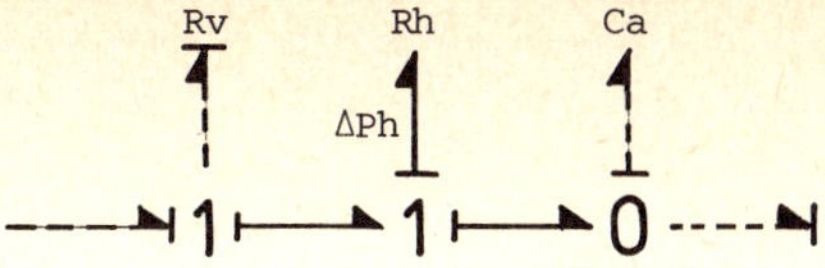

b. Including Hydraulic Line Capacitance (Ch)

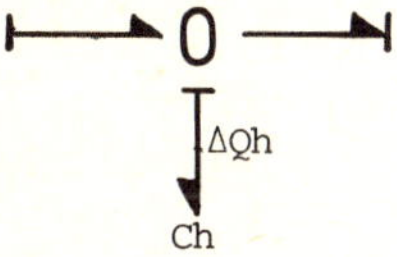

c. The Incompatibility of Showing Adjacent C Effects. (Where can the causal bar go on the centre bond?)

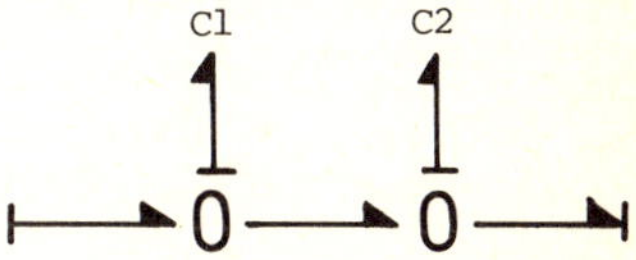

d. Resistance (Rh), Capacitance (Ch), Entry Resistance (Ra) in a Hydraulic Line Between a Control Valve (Rv) and Actuator (Ca)

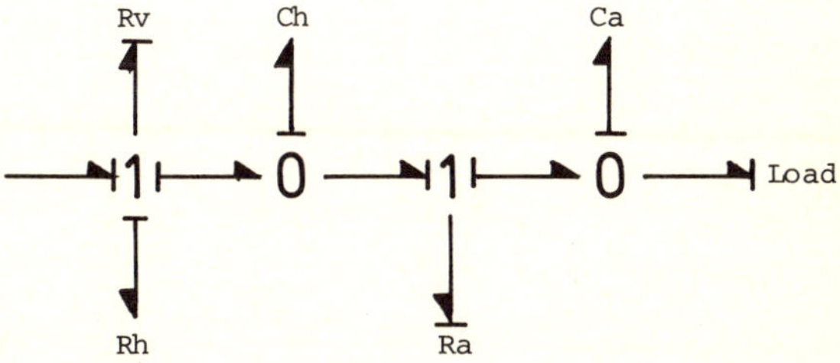

FIG. 7.20 BOND GRAPHS FOR HYDRAULIC LINES

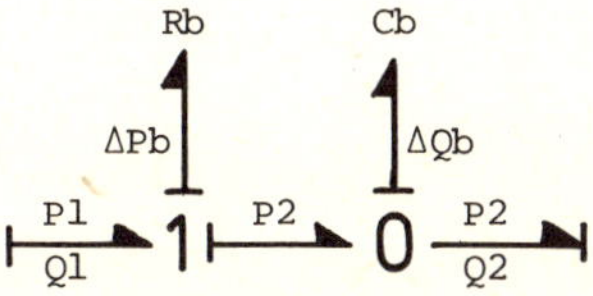

FIG. 7.21 ALLOWING FOR THE RESISTIVE (Rb) AND CAPACITIVE (Cb) EFFECTS OF A FILTER

a. Simple Capacitive Representation of an Accumulator

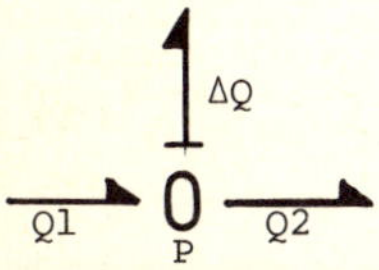

$$P = f(\Delta Q)$$
$$\Delta Q = Q1 - Q2$$

b. RC Representation of an Accumulator

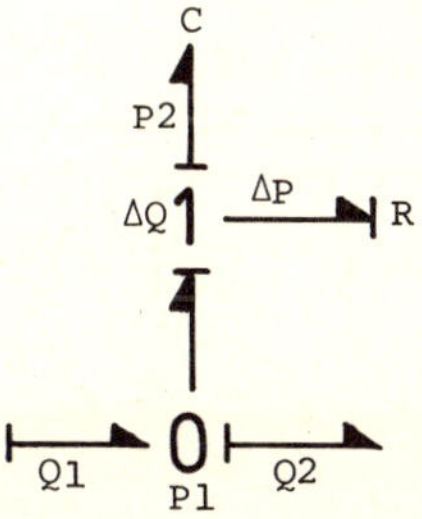

$$P2 = f(\Delta Q)$$
$$\Delta Q = f(\Delta P)$$
$$\Delta P = P1 - P2$$
$$Q2 = Q1 - \Delta Q$$
or $$Q1 = Q2 + \Delta Q$$

P1 and either Q1 or Q2 equations will arise from adjacent component bonds.

FIG. 7.22 BOND GRAPHS FOR ACCUMULATORS

effects of the line are represented as a series of RC effects. Fluid inertiance can be included in both the lumped parameter and the distributed parameter approaches. Multi-lump distributed parameter models allow for the time delay effect associated with long lines.

Accurate dynamic modelling of long fluid lines is a complex subject beyond the scope of the present Text. Ref. 7.7 provides a useful and relevant review.

7.11.2 Filters

A filter in a hydraulic line causes resistance (Rb) to flow which is reflected as a pressure drop, and a capacitive effect (Cb) due to its fixed volume. This RC effect is readily allowed for as the bond graph of Fig. 7.21 which implies the relationships

$$\begin{aligned} \Delta Pb &= f(Q1) \\ P1 &= P2 + \Delta Pb \\ P2 &= f(\Delta Qb) \\ \Delta Qb &= Q1 - Q2 \\ Q1 &= \text{(to be determined from upstream component)} \\ Q2 &= \text{(to be determined from downstream component)} \end{aligned} \qquad 7.30$$

Whether or not the Rb , Cb effect is significant enough to be allowed for in a particular system analysis must be decided by the analyst. Whether or not Rb can be lumped with an adjacent R , possibly of the hydraulic line, depends on the circumstances of the particular system and its bond graph. So too the Cb effect may be able to be lumped with an adjacent downstream C effect such as that of an actuator.

7.11.3 Accumulators

Accumulators are used in hydraulic control systems to provide some combination of the functions

- to reduce (damp out) flow and pressure pulsations in particular lines,
- to provide supply flowrate, and hence power additional to the pumps, for peak load demands of short duration,
- to maintain pressure in systems required to be inactive, but under pressure ready for use, for long periods. In these systems, the pump charges the accumulator as required and the accumulator supplies the system. The pump can be operated under automatic cut-in and cut-out control from pressure limit switches, or can be pressure-compensated if it needs.

to run continuously,

- to soften the pressure transients in systems having rapid stop-start or flow direction reversals.

Many hydraulic control systems do not have accumulators.

If an accumulator is simply providing additional volume to that of its associated line, it can be represented as the C effect illustrated on Fig. 7.22a. Note that the accumulator requires the flow-causal input bond (i.e. pressure is effect) and the pressure-causal output bond illustrated on Fig. 7.22a. Insertion of an accumulator in an existing system bond graph requires reassessment of causalities in the bond graph.

If, either in effect or by design, there is a significant resistance to flow into or from the accumulator, it behaves like the RC effect illustrated on Fig. 7.22b. The causal bars on the input (P1 · Q1) and output (P1 · Q2) bonds depend on the adjacent system components and could be the opposite to those shown.

The relevant functional relationships required have been added to Fig. 7.22. The C and R effects in no way imply linear relationships. However, unless there is reason to think them inappropriate, the expressions

$$P = 1/C \cdot \int Q \, dt + P(o) \qquad 7.31$$

where $C = V/B$

and $$\Delta Q = 1/R \cdot \Delta P \qquad 7.32$$

could be used. It should be borne in mind that V, the volume of oil under pressure in and around the accumulator, will vary depending on the state of charge of the accumulator. It should be borne in mind also that for most system analyses, the primary requirement will be to study the dynamics of the driven load, and not those of the accumulator. Usually a reasonable approximation of the accumulator dynamics is adequate. If a more accurate model of the accumulator dynamics is required, V may have to be appropriately modulated, and R may have to be described by a square root of pressure drop relationship.

Use of a large accumulator with the supply can justify an assumption of constant pressure supply to the control valve of a valve-controlled actuator system.

Accumulators and their characteristics are discussed in detail in Refs. 7.8 and 7.9.

7.11.4 Some Other Valves

Pressure relief valves and directional control valves have been discussed in Sections 7.6 and 7.9 respectively. It is not practicable to present here the bond graphs of the large number of valve variations available commercially. Several will be presented now, and it is left to the reader to form bond graphs for other valves when they are required.

Check Valve

The check valve, or non-return valve, is intended to allow free flow in one direction and to prevent totally flow in the other direction. Most check valves will have a flow-pressure drop characteristic of the kind idealized on Fig. 7.23a. Negative pressure drop results in no flow. A small positive pressure drop is usually present when flow is positive and this pressure drop can increase somewhat with flowrate. Pressure drop in the presence of flow means that there is power loss. Just how this is best accounted for on a bond graph depends on the location of the check valve relative to its adjacent components.

For example, consider the system of Fig. 7.23b. There will be three pressures involved when the load is being driven, these being supply Ps , pressure Px between control valve and check valve, and pressure Pa downstream of the check valve. To accommodate this the bond graph of Fig. 7.23c could be used. It includes capacitance Cx between control valve and check valve. The causality of the check valve Rx effect allows that flowrate through the check valve depends on the direction and size of the pressure drop ΔPx across it. Some appropriate equations are included with Fig. 7.23c.

Fig. 7.23d shows another interpretation in which the pressure characteristic of the check valve is taken as a source of pressure —— i.e. a pressure-effect input to the system which is substantially independent of the system. SPx would simply describe the characteristic of Fig. 7.23a. It is not always necessary to include check valves in limited dynamic analyses.

Counterbalance Valve

A counterbalance valve, Fig. 7.24a, is used to prevent an open-centre valve-controlled actuator with a suspended vertical load from running away when the valve is in neutral. Fig. 7.24b shows a typical application. The counterbalance valve allows free flow from the control valve to the underside of the actuator for the load-lifting stroke. It prevents flow from the actuator when the control valve is in neutral. In effect the counterbalance valve will not open to allow discharge flow until a certain set pressure is exceeded in the lower actuator line. This pressure is enough to generate a force on the underside of the actuator piston large enough to support the suspended load. If the control valve is moved to drive the actuator

a. Pressure-Flow Characteristic of Check Valve

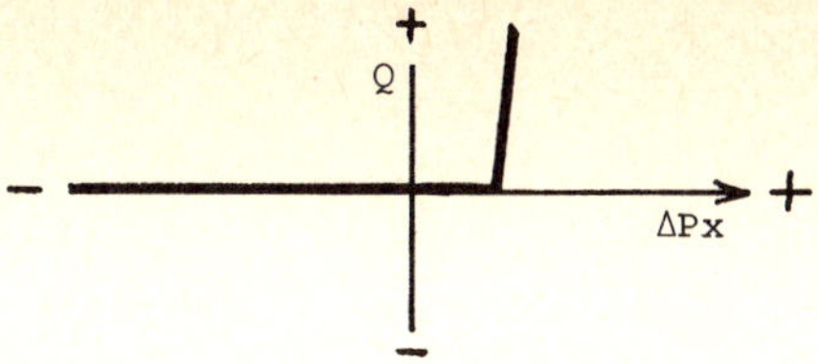

b. System with Check Valve

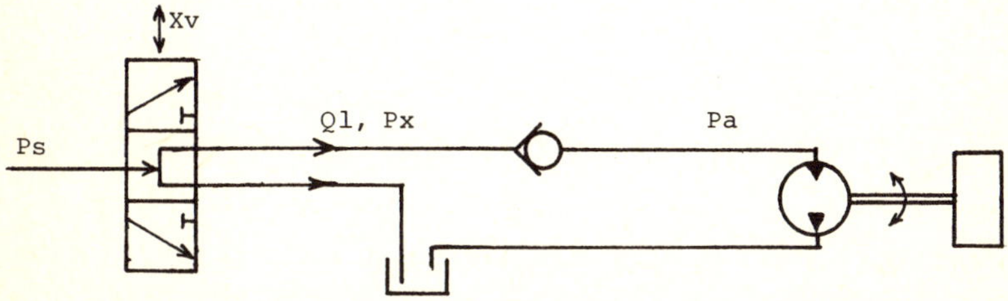

c. Bond Graph Allowing for Power Loss at the Check Valve

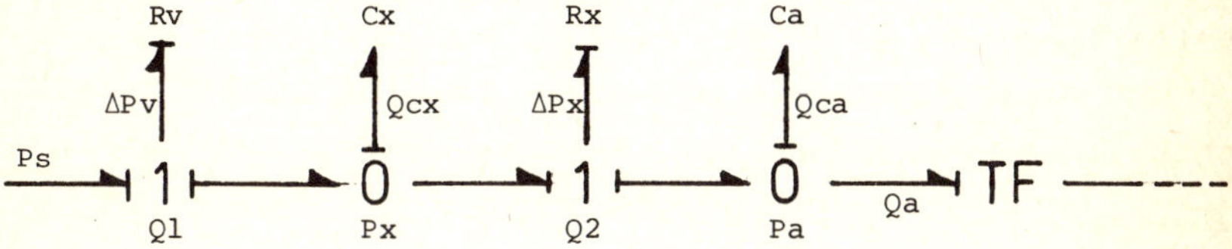

Q1	=	f(ΔPv, Xv)	ΔPv	=	Ps - Px
Px	=	f(Qcx)	Qcx	=	Q1 - Q2
Q2	=	f(ΔPx)	ΔPx	=	Px - Pa
Pa	=	f(Qca)	Qca	=	Q2 - Qa

d. Bond Graph when Check Valve Pressure Loss Taken as a Source

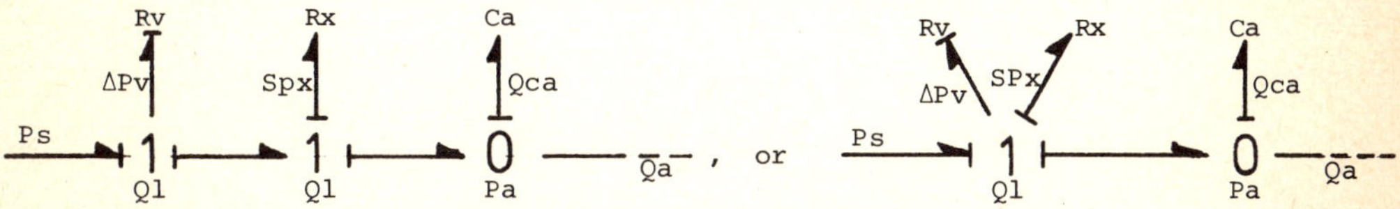

Q1 = f(ΔPv, Xv) ΔPv = Ps - SPx - Pa

Spx = specified from check valve characteristic

Pa = f(Qca), probably $1/Ca \cdot \int Qca \, dt + Pa(o)$

etc.

FIG. 7.23 INCLUDING CHECK VALVE ON BOND GRAPH

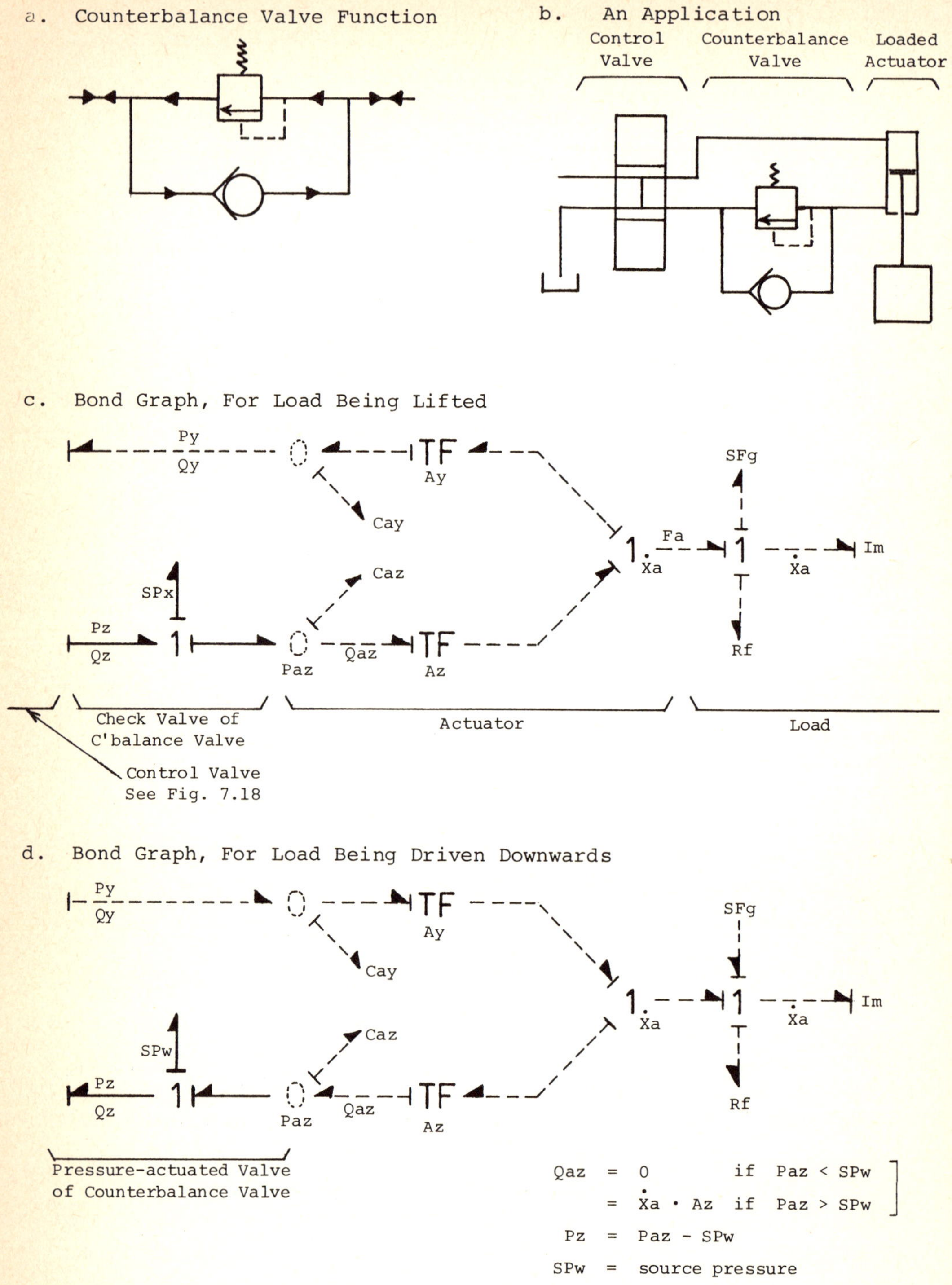

FIG. 7.24 INCLUDING A COUNTERBALANCE VALVE

down, the pressure build-up in the upper actuator line causes increased pressure in the lower side and the counterbalance valve opens to allow discharge flow, and hence downward motion of the load.

Fig. 7.24c shows a bond graph for the counterbalance valve for the condition that the load is being lifted. The check valve structure of Fig. 7.23d has been adopted. That is, the check valve is seen as providing a particular form of pressure source SPx . Supply Pz , Qz can be regarded as having come from the Z port of control valve structure of the form of Fig. 7.18, while discharge Py , Qy is feeding into the control valve's Y port. The various equations appropriate to Fig. 7.24c have all been developed previously.

Fig. 7.24d shows a bond graph for the counterbalance valve for the condition that the load is being driven downwards. The pressure-actuated valve is set at a certain threshold pressure. Thus, it can be shown on a bond graph with a symbol and causality similar to that used for a check valve. In this case the R effect is designated as the pressure source SPw . SPw will have a particular form which is substantially independent of the system in which the valve is used.

It is probably possible to combine Figs. 7.24c and d into a single bond graph. However, the algebra of the equations would become confusing. It is probably better to use the separate diagrams to prepare the joint equation set using the notions that Fig. 7.24c applies when control Xv is negative, and Fig. 7.24d when Xv is positive.

Counterbalance valves are sometimes pilot-operated by the pressure on the other side of the actuator. If this situation exists, the equations have to be appropriately modified.

7.12 CONCLUSION

The more common components, actions, and phenomena relating to the dynamic response of hydraulic control systems have been considered. The relationships adopted or proposed are not necessarily unique. A particular analyst may prefer different equation forms. Particular systems may require the use of additional equations or of different equations to describe adequately what the analyst feels will occur. With some experience in modelling, simulation, and in observation of hardware system performance, the designer-analyst can develop modified bond graphs and relationships. The objective of the present Chapter —— indeed of the present Text —— is to provide an approach to modelling, enough bond graph material, and enough equation forms, for simulation as a system design aid to be viable.

Quite apart from the deciding of equation forms, the selection of numerical values for coefficients in the equations is not always easy. Component manufacturers may have to be specifically solicited for some data.

7.13 REFERENCES

7.1 Merritt, H.E., Hydraulic Control Systems, Wiley, 1967.

7.2 Davis, D. and Stecki, J.S., Effective Bulk Modulus in Hydraulic Systems, Proc. Fluid Power Conf., FPS, Sydney, 1978.

7.3 Dransfield, P. and Bruce, D.M., Leakage Flow Rate Past Pistons of Oil Hydraulic System Components, J. of Aircraft, AIAA, V5, N2, 1968.

7.4 Barnard, B. and Dransfield P., Predicting Response of a Proposed Hydraulic Control System, Using Bond Graphs, Trans. ASME, JDSMC, V99, Series G, N1, 1977.

7.5 Chong, F.K. and Dransfield, P., The Effect of Choice of Relief Valve on the Response of a Hydraulic Control System, Proc. 1st Aust. Conf. on Control Eng., I.E. Aust., 1979.

7.6 Steber, G.R. and Romer, I.C., Simulation of a Pressure Relief Valve, Simulation, V12, N5, 1969.

7.7 Davis, D.C. and Stecki, J.S., Hydraulic System Analysis —— Modelling of Fluid Transmission Lines, BFPR Jrl., V14, N3, Oklahoma State University, 1980.

7.8 Knight, G.C., et al, Connection Capacitance Effects in Hydrostatic Transmission Systems and their Prediction by Mathematical Model, Proc. I.Mech.E., V186, 1972.

7.9 Svoboda, J., Analogue and Digital Modelling in the Design of a Hydraulic Vehicular Transmission, Proc. I.Mech.E., V193, 1979.

CHAPTER 8

APPLICATIONS OF BOND GRAPHS

8.1 INTRODUCTION

Chapters 4 and 5 introduced the ideas of power flow and power bond graph modelling respectively. Chapter 6 outlined the solution procedure once a model is obtained. Chapter 7 presented and discussed bond graph structures for components commonly used in hydraulic control systems. Application of the techniques will now be demonstrated more adequately using a series of case studies escalating in completeness and complexity. Where judged necessary, explanatory detail is provided.

8.2 CLOSED-CENTRE VALVE CONTROLLED INERTIA LOAD WITH FRICTION

Fig. 8.1a illustrates the system, which was used previously as an example in Chapter 5 (Section 5.5.4, Fig. 5.12). At that stage return line effects were neglected, the model was restricted to positive spool displacement only, and it was assumed that fluid was supplied to the valve at constant pressure. The example will now be extended to include return line effects and to allow spool displacement in both directions and thus load motion in either direction. The assumption of constant pressure supply will be retained, though its validity is questionable.

Fig. 8.1b shows the components considered to be relevant to the present analyis. Fig. 8.1c shows the bond graph, formed by direct coupling of the bond graphs of the major components, taken from Figs. 7.18 for the valve (with Rse neglected), 7.19b for the actuator (with Rfa and Ia deleted as they will be lumped with the load, the actuator and load being rigidly coupled; and internal leakage $Q\ell a$ neglected), and 7.5(2) for the load. Rsy , Rsz , Rye , and Rze have each been shown as Xv

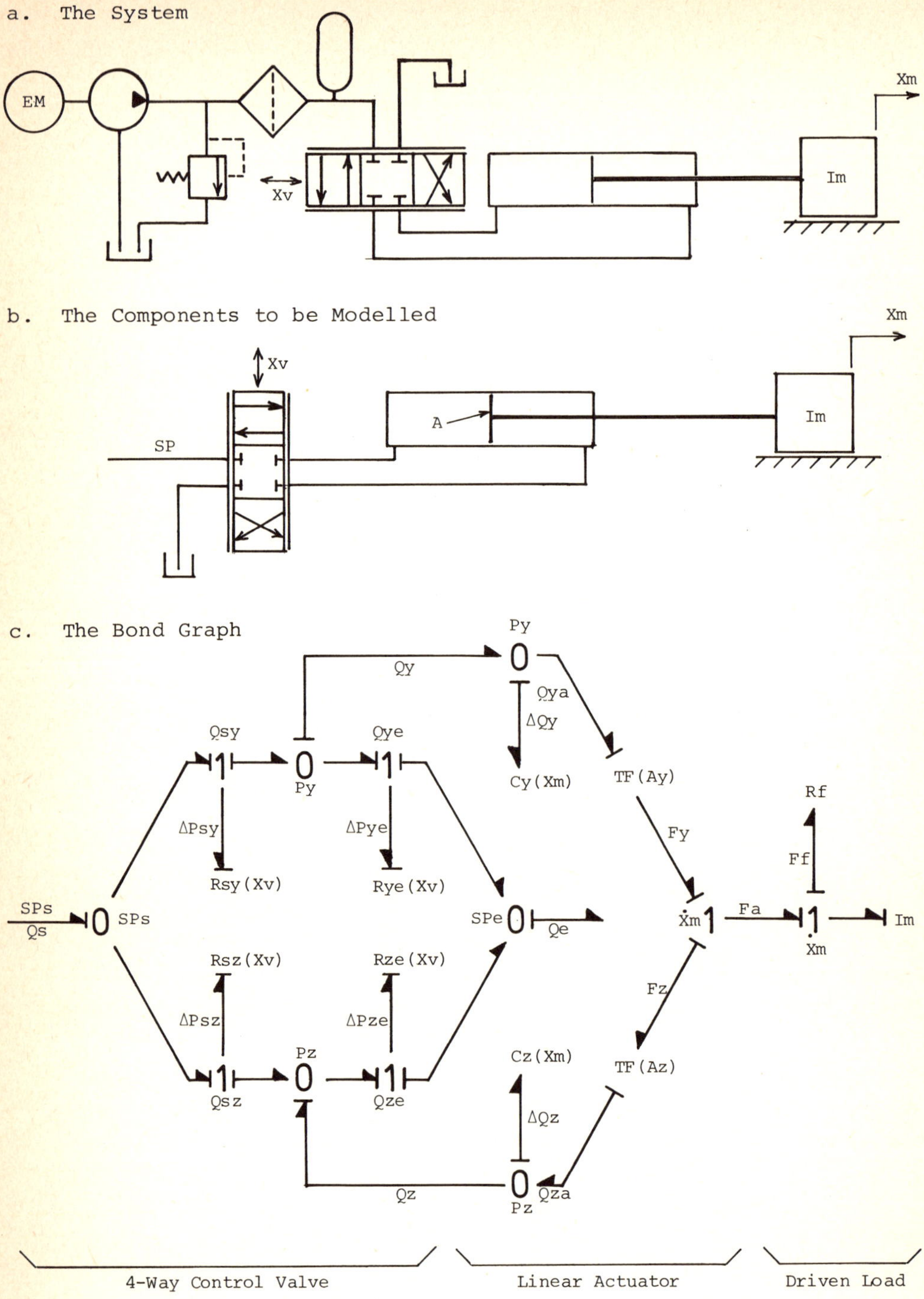

FIG. 8.1 SYSTEM OF EXAMPLE 8.1

dependent, and Cy and Cz each as Xm dependent. The Xm dependence of Cy and Cz could have been shown on the bond graph as broken lines with full arrowhead leading from $\dot{X}m$ to Cy and Cz , with integration of $\dot{X}m$ to show Xm .

Resistance of the hydraulic lines has been neglected or lumped with the valve resistance, and line capacitance has been neglected or lumped with actuator capacitance. The condition of flow into the Y end and from the Z end of the actuator has been taken as positive in assigning power flow directions. Each power state variable is shown once only on the bond graph. Using the notions that there can be only one effort variable on bonds around a particular 0-junction, and only one flow variable on bonds around a particular 1-junction, the reader can "see" the "missing" variables. Assignment of causality follows the patterns used previously in the Text.

Having decided that the bond graph structure of Fig. 8.1c adequately reflects the dynamic effects required in the model, it is required next to choose an equation for each power state variable. The equations, developed in the order Sources, I effects, C effects, R effects, Transformations, Junctions, are given in Table 8.1.

The first 26 equations are the power state equations. Bond graph flow directions, and the equations, were prepared on the basis that Xv would be positive. The equations hold equally for Xv negative. In that case Qy , Qya , Fy , and Qz , Qza , Fz each become negative relative to their directions assigned on the bond graph.

The model is completed by adding the four auxiliary equations 8.1.27 to 30. To solve (simulate) the model, values are required for initial values $\dot{X}m(o)$, Py(o) , Pz(o) , Xm(o) , and for coefficients SPs , SPe , Im , Kv , Fμ , Kf , Ay , Az , Vy(o) , Vz(o) , B and the input Xv(t) needs to be specified. The simulation would provide system response to the specified input. It is likely that the analyst would arrange for plots of Xm(t) , Py(t) , Pz(t) , and possible Qsy(t) and Qsz(t) , although other or more variables can be studied if required.

If the control valve had overlap, or if it was of open-centre or tandem-centre configuration, the equations are readily adjusted using the details given in Section 7.9. The valve flow equations 8.1.6, 7, 8, and 9 can be readily expanded to allow for saturation of port opening, also as demonstrated in Section 7.9.

8.3 LOADED HYDRAULIC SERVOSYSTEM

Fig. 8.2a illustrates a type of servosystem in wide use in proportional positioning systems (machine tool slides, aircraft control surfaces, etc.). Assuming constant pressure supply, an assumption strengthened by the presence of the accumulator upstream of the control valve, the components essential to the dynamic analysis are shown on Fig. 8.2b. This configuration was modelled conventionally in Chapter 3 without the presence of the external force Fe now included as part of the load.

Fig. 8.2c shows the bond graph proposed for the system. It is similar to Fig. 8.1c except for

- the allowance for compliant restraint of the load (force Fk on the Ck bond)
- allowance for the external force Fe (SFe bond)
- combining of the adjacent Fa and Fm 1-junctions
- indication that the valve resistances are both Xi (input) and Xm (load displacement) dependent. Spool displacement Xv will be both Xi and Xm dependent
- Cy and Cz being shown as Xm dependent via direct feedback of $\dot{X}m$.

The set of equations is similar to the 8.1 set with the modifications

- the actuator-to-load 1-junction(s) equations 8.1.25 and 8.1.26 become

$$Fm = Fy - Fz - Ff - Fk - SFe \qquad 8.2.1$$

- the Xv equation 8.1.30 becomes

$$Xv = (L2 - L1)/L2 \cdot Xi - L1/L2 \cdot Xm \qquad 8.2.2$$

- the input equation becomes

$$Xi = \text{to be specified} \qquad 8.2.3$$

- additional load force equations are

$$SFe = \text{constant} \qquad 8.2.4$$

$$Fk = 1/Ck \cdot \int \dot{X}m \, dt + Fk(o) \qquad 8.2.5$$

8.4 PUMP SUB-SYSTEM

Most hydraulic control systems require a pumping unit (power supply) consisting of

- the primary drive (electric motor, diesel engine, aircraft engine, etc.)
- a positive displacement pump (fixed or variable delivery)

TABLE 8.1 Model for Example 8.1

Component, Effect, Comment	*Equation*	
Power State Equations:		
Supply, source	$SPs = \text{constant}$	8.1.1
Exhaust, source	$SPe = \text{constant}$	8.1.2
Im effect	$\dot{Xm} = 1/Im \cdot \int Fm\,dt + \dot{Xm}(o)$	8.1.3
Cy effect; Xm dependent cap. of y side of actuator	$Py = 1/Cy \cdot \int \Delta Qy\,dt + Py(o)$	8.1.4
Cz effect; Xm dependent cap. of z side of actuator	$Pz = 1/Cz \cdot \int \Delta Qz\,dt + Pz(o)$	8.1.5
Rsy effect; Xv dependent $Qsy = f(\Delta Psy)$ No saturation of port opening	$Qsy = 0$ if Xv is 0; $= Kv \cdot Xv \cdot \Delta Psy^{\frac{1}{2}}$ if Xv is positive; $= 0$ if Xv is negative	8.1.6
Rsz effect	$Qsz = 0$ if Xv is 0; $= 0$ if Xv is positive; $= Kv \cdot \lvert Xv\rvert \cdot \Delta Psz^{\frac{1}{2}}$ if Xv is negative	8.1.7
Rye effect	$Qye = 0$ if Xv is 0; $= 0$ if Xv is positive; $= Kv \cdot \lvert Xv\rvert \cdot \Delta Pye^{\frac{1}{2}}$ if Xv is negative	8.1.8
Rze effect	$Qze = 0$ if Xv is 0; $= Kv \cdot Xv \cdot \Delta Pze^{\frac{1}{2}}$ if Xv is positive; $= 0$ if Xv is negative	8.1.9
Load friction, Rf effect	$Ff = \text{Sign}\,\dot{Xm} \cdot F\mu + Kf \cdot \dot{Xm}$	8.1.10
Power transformation, TF(Ay)	$Qya = \dot{Xm} \cdot Ay$	8.1.11
	$Fy = Py \cdot Ay$	8.1.12
Power transformation, TF(Az)	$Qza = \dot{Xm} \cdot Az$	8.1.13
	$Fz = Pz \cdot Az$	8.1.14
Supply 0-junction	$Qs = Qsy + Qsz$	8.1.15
Rsy 1-junction	$\Delta Psy = SPs - Py$	8.1.16
Qy 0-junction	$Qy = Qsy - Qye$	8.1.17
Rye 1-junction	$\Delta Pye = Py - Pe$	8.1.18
Exhaust 0-junction	$Qe = Qye + Qze$	8.1.19
Rze 1-junction	$\Delta Pze = Pz - Pe$	8.1.20
Qz 0-junction	$Qz = -Qsz + Qze$	8.1.21
Rsz 1-junction	$\Delta Psz = SPs - Pz$	8.1.22
Cy 0-junction	$\Delta Qy = Qy - Qya$	8.1.23
Cz 0-junction	$\Delta Qz = Qza - Qz$	8.1.24
Fa 1-junction	$Fa = Fy - Fz$	8.1.25
Fm 1-junction	$Fm = Fa - Ff$	8.1.26
Auxiliary Equations:		
Load displacement Xm(t)	$Xm = \int \dot{Xm}\,dt + Xm(o)$	8.1.27
Vy(o) is initial volume, Y side, B is bulk modulus	$Cy = \left(Vy(o) + Ay \cdot Xm\right)/B$	8.1.28
Vz(o) is initial volume, Z side, B is bulk modulus	$Cz = \left(Vz(o) + Az \cdot Xm\right)/B$	8.1.29
	$Xv = \text{input to be specified}$	8.1.30

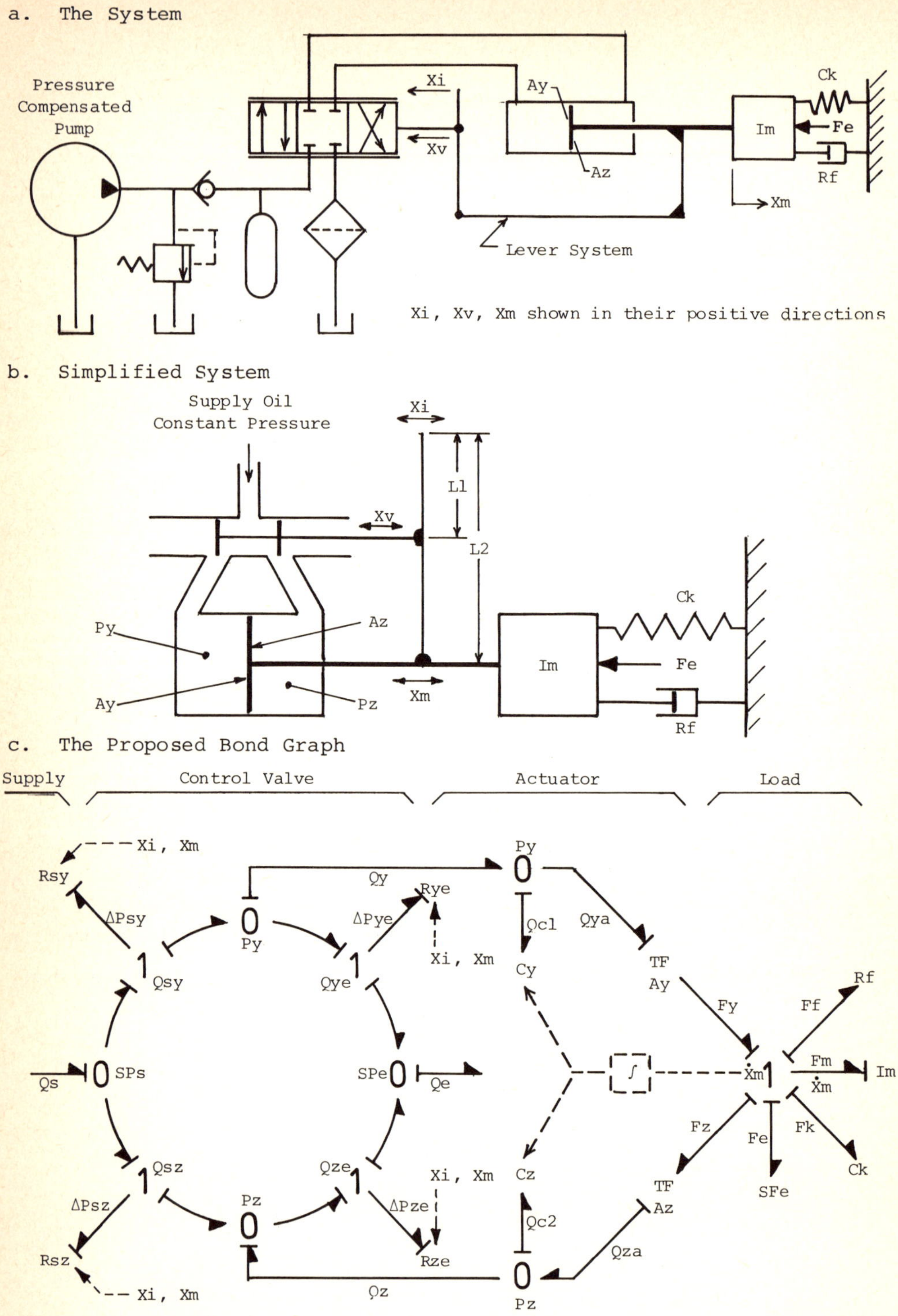

FIG. 8.2 BOND GRAPH FOR LOADED HYDRAULIC SERVOSYSTEM

. a pressure relief valve

. a reservoir and its ancillary equipment

Many systems also include a high pressure filter near the pump discharge. Some include an accumulator. It is reasonable to think of a power unit as being versatile in that it could be used to provide power to any of a number of actuation systems. Hence, for the system design situation it is attractive to have on hand dynamic models for typical power supply sub-systems.

Fig. 8.3a shows one typical pumping system. The variable restriction is included to represent the impedance load the sub-system will "see" when it is connected to an actuator sub-system. The intention now is to prepare a fairly detailed model which could be useful in a variety of applications of the pumping system. Such a model could also be used to study the internal dynamics of the pumping system itself.

The system can be conveniently divided into the three sections

. the pump drive sub-system (electric motor, shafting, coupling)

. the pump

. the downstream components (relief valve, line, filter, valve).

Previous Figs. 5.4, 5.8, 7.9 developed a bond graph for an induction electric motor, including its output shaft compliance. Fig. 7.9 is repeated as Fig. 8.3b. The coupling and pump shaft unit can be included by lumping its inertia with rotor inertia Ix , its friction with Rf and its compliance with Ck . From previous Fig. 7.11 the bond graph of Fig. 8.3c can be proposed for the pump. Fig. 8.3d shows a bond graph encompassing the relief valve (Rr) , the resistance of the line plus filter (Rb) , the capacitance of the line plus filter (Cb) , and the resistance of the load valve (Rv) .

Fig. 8.3e shows the system bond graph formed simply by coupling the bond graphs of Figs. 8.3b, c and d. Adjacent 1-junctions and adjacent 0-junctions have been combined. The direction half-arrows and the causal bars were readily added using physical sense and causal preferences. Combining the component sub-systems did not upset their power flow directions and causalities.

The system equations written using the bond graph are given in Table 8.2. The terms and symbols are as previously used.

Linear relationships were assumed for most of the dynamic effects in the model. This assumption allows effects such as Ix , Ip , Ck , Cb , $R\omega$, $R\ell p$, Rr , Rb , and Rv to be taken as linear coefficients. This is by no means always necessary or desirable. It is obvious enough that Ix and Ip are coefficients —— the respective inertia values. Shaft compliance Ck is reasonably a coefficient. The capacitance Cb is associated in the present example with fixed volumes, and consequently the assumptions of linearity is reasonable. Linearity of Rω is reasonable over a fairly wide range of electric motor performance. Leakage flow through the clearances of the pump's working parts is viscous, and hence linear with pressure over wide ranges of pressure. Assumption that the resistance Rb of the filter and hydraulic line is linear is more suspect if we expect the model to give large scale dynamic response. However, the pressure drop associated with Rb is quite small relative to the pressure drop likely over the adjacent load valve Rv . Consequently we probably need not describe Rb with much accuracy.

The overall load pressure drop, represented by ΔPv across resistance Rv , is undoubtedly large, and consequently we describe it by a normal square law orifice flow relationship (equation 8.3.13).

The friction equations used for Tf and Tfp were discussed in Chapter 7. Coulomb friction could be added.

The TF equations are the well known ideal relationships

pump torque = volumetric displacement per radian (Vp) · pump pressure (Pp)

pump flow = volumetric displacement per radian (Vp) · pump speed (ωp)

For a variable displacement pump, volumetric displacement per radian of rotation per radian of swashplate angle (Vω) multiplied by the swashplate angle (θp) in radians gives the required pump volumetric displacement (equation 8.3.24).

The model, equations 8.3.1 to 25, was applied to a specific system, which was also built and tested experimentally (Refs. 8.1, 8.2). A fixed restriction Rv was used, and the system was activated for dynamic response by moving swashplate angle θp from one position to another, increasing the flowrate demand of the system.

Some of the coefficients needed for the model were easily obtained, others required experimental investigation or estimation. The values and how they were obtained are indicated in Table 8.3.

Fig. 8.4 shows the results of a simulation of the model effected using CSMP on a B6700 computer. Fig. 8.4a shows the input used, a change in swashplate angle. Fig. 8.4b shows the consequent development of system pressure Pv(t) , together with a corresponding experimental measurement. Fig. 8.4c shows simulation results using a variety of values for the bulk modulus of the oil. Once a simulation is working it is a simple matter to make this sort of excursion to investigate the sensitivity

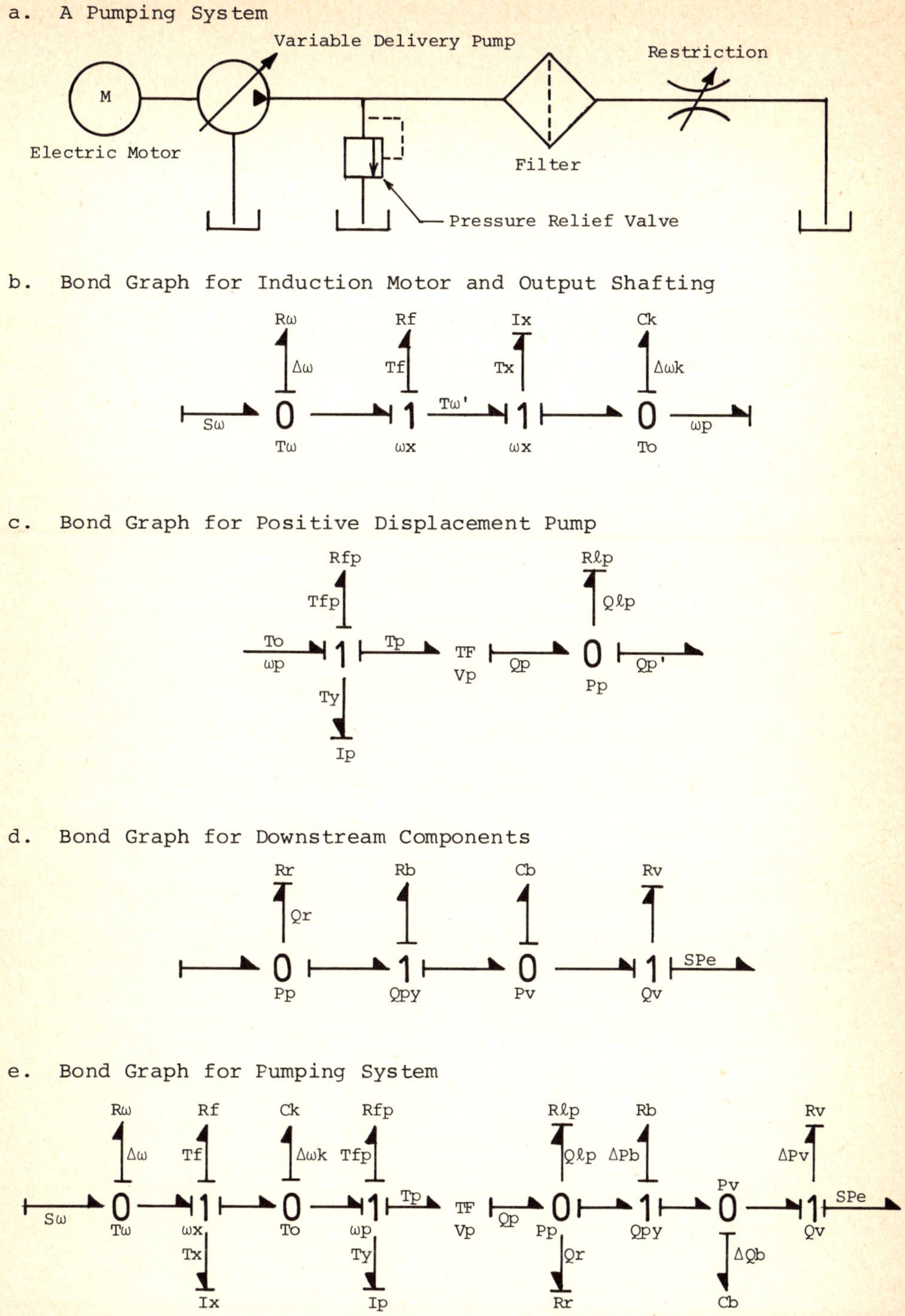

FIG. 8.3 BOND GRAPH FOR PUMPING SYSTEM (EXAMPLE 8.3)

TABLE 8.2 Power State Equations for Example 8.4

Component, Effect, Comment	*Equation*	
Sources:		
Electric motor field speed	$S\omega = \text{constant}$	8.3.1
Reservoir pressure	$Spe = 0$	8.3.2
I effects:		
Electric motor shaft speed	$\omega x = 1/Ix \cdot \int Tx\, dt + \omega x(o)$	8.3.3
Pump input speed	$\omega p = 1/Ip \cdot \int Ty\, dt + \omega p(o)$	8.3.4
C effects:		
Electric motor output torque	$To = 1/Ck \cdot \int \Delta\omega k\, dt + To(o)$	8.3.5
System discharge pressure	$Pv = 1/Cb \cdot \int \Delta Qb\, dt + Pv(o)$	8.3.6
R effects:		
Motor torque, $R\omega$ effect	$T\omega = R\omega \cdot \Delta\omega$	8.3.7
Motor windage plus motor shaft friction, Rf effect	$Tf = Kf1 \cdot \omega x^3 + Kf2 \cdot \omega x$	8.3.8
Pump friction, Rfp effect	$Tfp = Kf3 \cdot \omega p$	8.3.9
Pump leakage, $R\ell p$ effect	$Q\ell p = 1/R\ell p \cdot Pp$	8.3.10
Relief valve, Rr effect	$Qr = 0$ if $Pp < Pset$ $= 1/Rr \cdot (Pp - Pset)$ if $Pp > Pset$	8.3.11
Line pressure drop, Rb effect	$\Delta Pb = Rb \cdot Qpy$	8.3.12
Discharge flowrate, Rv effect	$Qv = 1/Rv \cdot \Delta Pv^{\frac{1}{2}}$	8.3.13
TF:		
Pump input torque	$Tp = Vp \cdot Pp$	8.3.14
Pump geometric discharge rate	$Qp = Vp \cdot \omega p$	8.3.15
Junctions:	$\Delta\omega = S\omega - \omega x$ (motor speed slip)	8.3.16
	$Tx = T\omega - Tf - To$	8.3.17
	$\Delta\omega k = \omega x - \omega p$ (shaft torsional deflection)	8.3.18
	$Ty = To - Tfp - Tp$	8.3.19
	$Qpy = Qp - Q\ell p - Qv$	8.3.20
	$Pp = Pv + \Delta Pb$	8.3.21
	$\Delta Qb = Qpy - Qv$ (compressibility component)	8.3.22
	$\Delta Pv = Pv - Spe$	8.3.23
Auxiliary Equations:	$Vp = V\omega \cdot \theta p$	8.3.24
	θp = controlling input, swashplate angle	8.3.25

TABLE 8.3 Coefficients for Example 8.4

Coefficient		*Value*		*Dimension*	*Source*
$S\omega$	=	157		rad/s	1500 rpm syn. field speed (manufacturer)
Ix	=	16.5	E-3	$kg \cdot m^2$	rotor inertia (manufacturer) plus shaft and coupling inertia (evaluated)
Ip	=	1.15	E-3	$kg \cdot m^2$	pump rotor inertia (manufacturer)
Ck	=	10	E-6	$rad/(N \cdot m)$	length/(M of I · shear mod.) of shaft
Cb	=	3.3	E-12	m^3/Pa	Vol. 4.52 E-3 m^3/bulk mod. 1380 MPa
$R\omega$	=	53		$N \cdot m/(rad/s)$	slip coefficient (manufacturer)
Kf1	=	0.316	E-3	$N \cdot m/(rad/s)^3$	estimated rotor speed friction coefficient
Kf2	=	3.2	E-9	$N \cdot m/Pa$	estimated shaft viscous friction coef.
Kf3	=	5	E-9	$N \cdot m/Pa$	estimated pump rotor friction coefficient
$R\ell p$	=	0.5	E12	$Pa/m^3/s)$	pump leakage coefficient (manufacturer)
Rr	=	4.1	E9	$Pa/(m^3/s)$	relief valve characteristic (manufacturer)
Pset	=	28	E6	Pa	analyst's decision
Rb	=	0.91	E9	$Pa/m^3/s)$	manufacturer
Rv	=	8.9	E9	$Pa^{\frac{1}{2}}/m^3/s)$	from experimental $\Delta Pv - Qv$ plot
$V\omega$	=	19.7	E-6	$(m^3/rad)/rad$	pump manufacturer's data

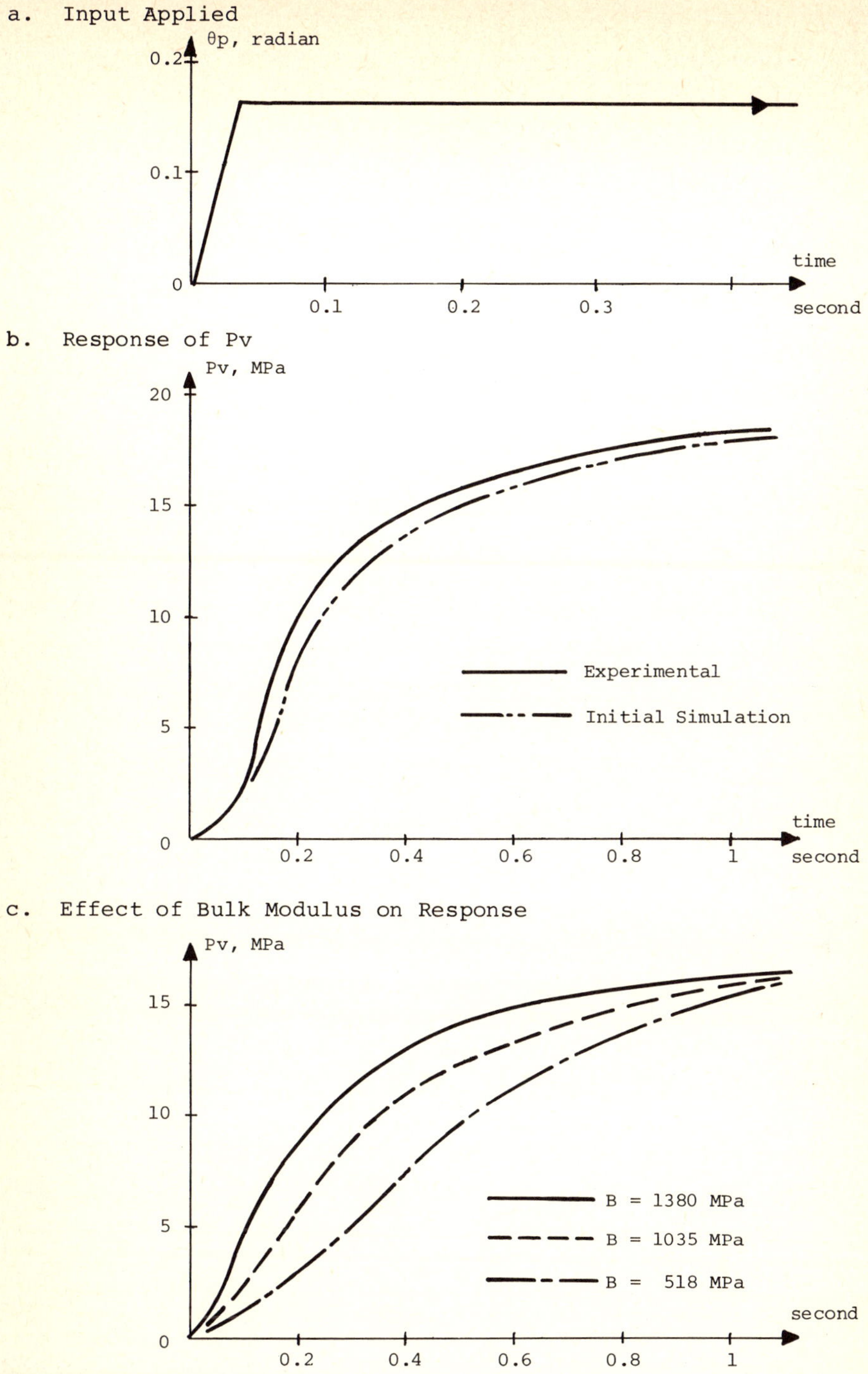

FIG. 8.4 SIMULATION OF MODEL FOR EXAMPLE 8.3

of a system to changes in specific parameters. The dynamic development of all system variables was available.

It is probable that if it was required only to study Pv(t) as θp is varied, the bond graph and its equation set could be simplified greatly. Pump speed could probably be taken as a source, allowing dismissal of the pump drive's dynamics and of pump friction and inertia. It is probable that line resistance Rb also could be ignored as small compared with Rv . However, if detailed study of the pump unit was required, the detailed model should be used.

8.5 PUMP-CONTROLLED HYDROSTATIC DRIVE

Fig. 8.5a illustrates the system to be simulated. It is assumed that

- the speed of the input drive is constant
- load speed is controlled by external adjustment Xi of the pump stroke control lever
- pump displacement can be zero or positive only
- the hydraulic circuit remains filled so that there is no cavitation (make-up components not shown)
- the relief valve will not normally lift, being purely a safety device.

A version of this system was used as an example for conventional modelling in Chapter 3.

Fig. 8.5b shows a bond graph for the system. It utilizes a structure for the pump modified from Fig. 7.11g by neglecting pump friction (Rfp) and inertia (Ip) . It includes delivery line resistance Rh and hydraulic motor leakage $R\ell a$. It neglects the effects on dynamic response of the return line flow. Pump, delivery line, and motor capacitances are lumped as Cpha . The hydraulic motor is regarded as a fixed-modulus power transformer, while the driven assembly is represented as a rotary inertia load with friction. The set of power state equations prepared using the bond graph is shown on Table 8.4, the nomenclature being included with Fig. 8.4.

The initial values $\omega j(o)$, Pp(o) and the coefficients $S\omega p$, Ij , Vpha , B , $K\ell p$, Kr , Kh , $K\ell a$, Kfj , Va , Kp must be specified before simulation can proceed. The simulation then provides $\omega j(t)$, Pp(t) , etc. for specific inputs Xi(t) . The model and simulation can be extended to allow for the effects of the return line, for cross-port leakage, and for over-centre pump action (reversible load drive).

a. Schematic of System

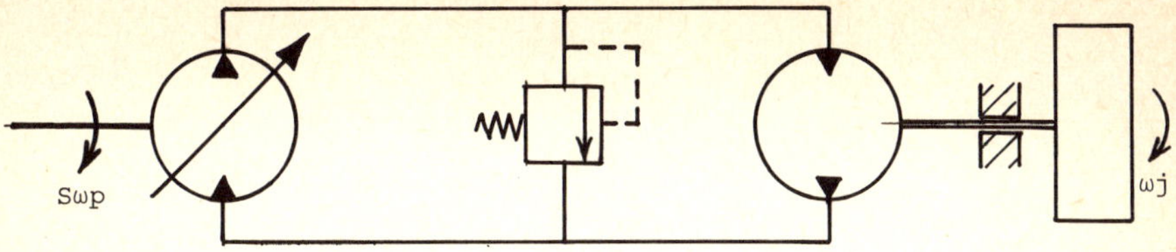

b. Proposed Bond Graph

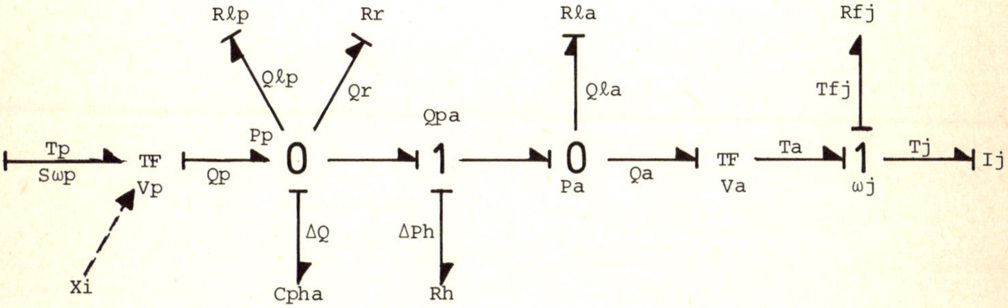

c. Definition of Symbols

Sωp	constant angular velocity of input drive	Qpa	flowrate into motor
Tp	torque developed in input drive	Qℓa	leakage in motor
Vp	swept volume of pump per radian rotation	Pa	pressure at motor
Xi	angular displacement of swashplate	Qa	swept volume per second of motor
Qp	swept volume of pump per second	Va	swept volume per radian of motor
Pp	pressure developed by pump	Ta	torque developed by motor
Qℓp	leakage in pump	Tfj	friction torque in motor per load
Qr	flowrate through relief valve	ωj	angular speed of motor per load
ΔQ	loss of flowrate due to oil compression	Tj	torque accelerating load
ΔPh	pressure drop in line	Vpha	volume under compression
		B	bulk modulus of oil
		K	coefficient for an R effect

FIG. 8.5 PUMP CONTROLLED HYDROSTATIC DRIVE

TABLE 8.4 Model for Example 8.5

Effect		*Equation*	
Power State Equations:			
Supply source	$S\omega p$	$= \text{constant}$	8.4.1
Driven inertia Ij	ωj	$= 1/Ij \cdot \int Tj\, dt + \omega j(o)$	8.4.2
Hydraulic capacitance Cpha	Pp	$= 1/Cpha \cdot \int \Delta Q\, dt + Pp(o)$	8.4.3
Pump leakage $R\ell p$	$Q\ell p$	$= K\ell p \cdot Pp$	8.4.4
Relief valve Rr	Qr	$= 0$ if $Pp < Pset$ $= Kr \cdot (Pp - Pset)$ if $Pp > Pset$	8.4.5
Line resistance Rh	ΔPh	$= Kh \cdot Qpa$	8.4.6
Motor leakage $R\ell a$	$Q\ell a$	$= K\ell a \cdot Pa$	8.4.7
Load friction Rfj	Tfj	$= Kfj \cdot \omega j$	8.4.8
Pump transformer Vp	Tp	$= Vp \cdot Pp$	8.4.9
	Qp	$= Vp \cdot S\omega p$	8.4.10
Motor transformer Va	Ta	$= Va \cdot Pa$	8.4.11
	Qa	$= Va \cdot \omega j$	8.4.12
Pump 0-junction	ΔQ	$= Qp - Q\ell p - Qr$	8.4.13
Line 1-junction Rh	Pa	$= Pp - \Delta Ph$	8.4.14
Motor 0-junction $R\ell a$	Qpa	$= Q\ell a + Qa$	8.4.15
Load 1-junction Tfj	Tj	$= Ta - Tfj$	8.4.16
Auxiliary Equations:			
Capacitance of pump, hose and actuator	$Cpha$	$= Vpha/B$	8.4.17
Pump displacement/rad	Vp	$= Kp \cdot Xi$	8.4.18
Control input	Xi	$=$ input to be specified	8.4.19

8.6 VALVE-CONTROLLED HYDROSTATIC DRIVE

The system, illustrated on Fig. 8.6a, was used to demonstrate power port modelling in Chapter 4. It is developed from Ref. 8.3. Fig. 8.6b repeats the power port schematic so that the reader can easily study the relationships between the system, its power port schematic, and its bond graph which is shown on Fig. 8.6c.

The bond graph implies a quite detailed dynamic model, as it includes

- a pressure compensated pump with its input drive
- a relief valve
- a filter
- a 4-way directional control valve
- a cross-port relief valve system, with make-up supply from the charge pump
- a hydraulic motor
- the driven rotary inertia load.

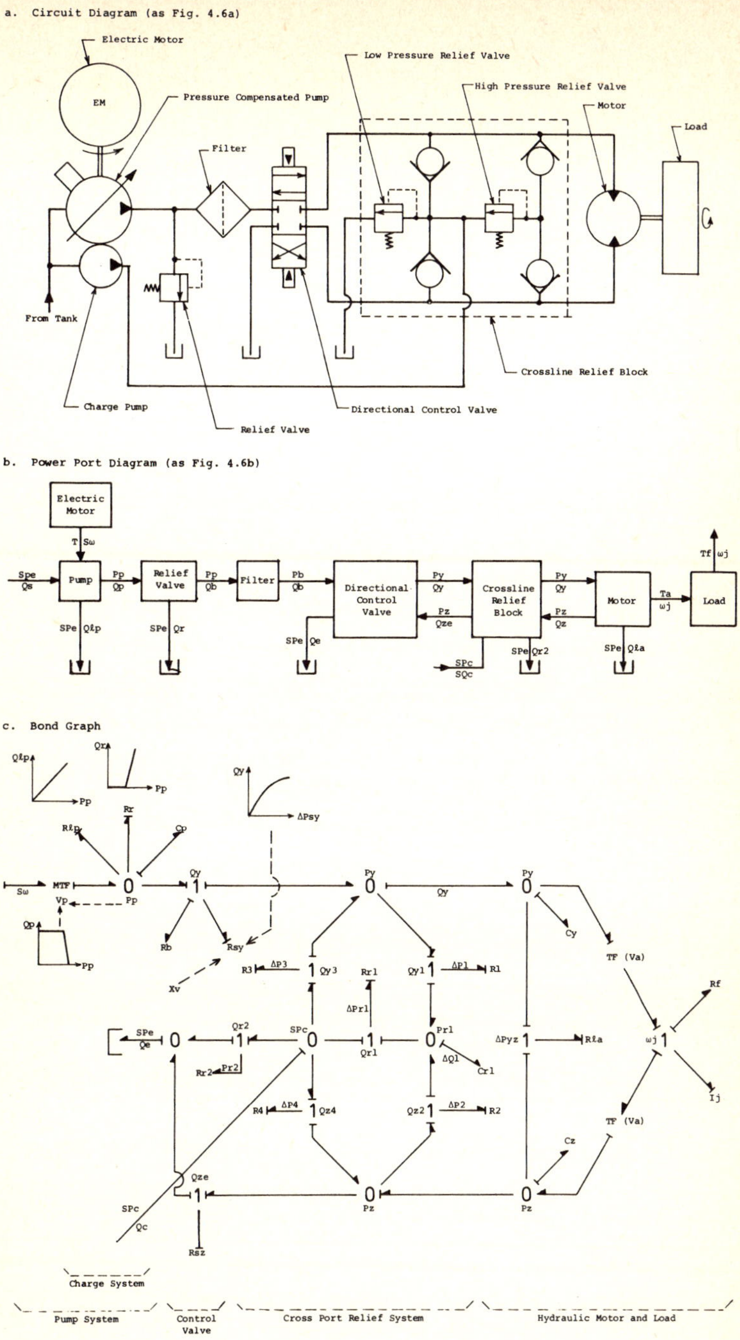

FIG. 8.6 VALVE-CONTROLLED HYDROSTATIC DRIVE (EXAMPLE 8.6, TABLE 8.5)

The additional information and model development on a bond graph relative to that on a power port schematic is clearly illustrated. Each has a block representing the system's components or sub-systems. Each shows the main power flow bonds. But that is all that the power port schematic shows, while the bond graph goes on to expose those effects in each block which are necessary, or that the analyst decides are significant, to the system's dynamic performance. Further, the decisions on causality provide both a technique for checking the appropriateness of a developing model, and a framework for arrangement of the equations. The full arrows on the power port diagram and the half arrows on the bond graph both indicate the presumed directions of power flow in the system, and provide equation sign conventions.

The common effort at an 0-junction or flow at a 1-junction is shown adjacent to the junction so that it need not be repeated on each bond of the particular junction. Information on the present bond graph has been further advanced by inclusion of graphic symbols indicating the nature of the non-linear relationships present. Thus, the symbol at the pump MTF shows clearly that the pump is pressure compensated. The graphic symbols are shown connected to the bond graph by active bonds (broken lines, full arrowheads). Note the feedback from the pump discharge side 0-junction to the pressure compensator, shown by an active bond. The 4-way control valve bond graph allows only for positive load rotation. This decision reduces the valve's bond graph from that of Figs. 8.1 and 8.2 to the two 1-junction resistances Rsy and Rze , as Rsz and Rye are not allowed for. If required, full bi-directional valve (and hence load) actuation can be reinstated.

Equations for most of the system's components have been presented in previous examples and will not be repeated here. The cross-port relief block has not been detailed previously. Using the bond graph its actions can be described by the equations of Table 8.5.

The mathematical description, comprised of equations 8.5.1 to 19, is quite extensive. But so too is the cross-port relief system itself, comprising of four non-return (check) valves, two pressure relief valves and the various interconnections between these and with the system itself. The bond graph and the equations given allow that the cross-port relief system might have to work in either direction of motion of the hydraulic motor and load. The bond graph structure of the control valve on Fig. 8.6c allows that the motor-load rotation will be in one direction only. However, even with this restriction on control valve actuation, the cross-port relief system may have to act in either of its directions.

TABLE 8.5 Equations for Cross-Port Relief Block, Example 8.6

Effect	*Equation*		
Flow through check valve R1 from Y line, and the associated 0- and 1-junctions	$Qy1 = 0$ $= K1 \cdot \Delta P1$	if $\Delta P1$ positive if $\Delta P1$ negative	8.5.1
	$\Delta P1 = Py - Pr1$		8.5.2
	$Qy = Qy - Qy1 + Qy3$		8.5.3
Flow through check valve R2 from Z line, and the associated 0- and 1-junctions	$Qz = Qz2 + Qze - Qz4$		8.5.4
	$Qz2 = 0$ $= K2 \cdot \Delta P2$	if $Pz < Pr1$ if $Pz > Pr1$	8.5.5
	$\Delta P2 = Pz - Pr1$		8.5.6
Pressure development and flow downstream of check valves 1 and 2 into capacitance Cr1	$Pr1 = 1/Cr1 \cdot \int \Delta Q1\, dt + Pr1(o)$		8.5.7
	$\Delta Q1 = Qy1 + Qz2 - Qr1$		8.5.8
Flow through high pressure relief valve Rr1	$Qr1 = 0$ $= Kr1 \cdot (Pr1 - P1\ set)$	if $Pr1 < P1\ set$ if $Pr1 > P1\ set$	8.5.9
	$\Delta Pr1 = Pr1 - SPc$		8.5.10
Charging supply; assumed to be at constant pressure and flowrate, SPc and SQc ; no need then to include equations for relief valve Rr2	$SPc = constant = P2\ set$		8.5.11
	$SQc = constant$		8.5.12
	$Qr2 = SQc - Qr1 - Qy3 - Qz4$		8.5.13
	$\Delta Pr2 = SPc - SPe = constant$		8.5.14
	$SPe = constant$		8.5.15
Flow through check valve R3 into Y line, and the associated 1- and 0-junctions	$Qy3 = 0$ $= K3 \cdot \Delta P3$	if $Py \geqslant Pc$ if $Py < Pc$	8.5.16
	$\Delta P3 = SPc - Py$		8.5.17
Flow through check valve R4 into Z line, and the associated 1- and 0-junctions	$Qz4 = 0$ $= K4 \cdot \Delta P4$	if $Pz \geqslant Pc$ if $Pz < Pc$	8.5.18
	$\Delta P4 = SPc - Pz$		8.5.19

8.7 A LIFTING SYSTEM

8.7.1 Introduction

The requirement was to position a load in a vertical arc, using a hydraulic cylinder to actuate a pivoted beam, as illustrated on Fig. 8.7a. The specifications were

- the load is 125 kg;
- full load travel is 0.6 m;
- control is to be manual;
- the system is to be capable of causing the load to change position by 0.3 m in 1 second;
- load acceleration following the input command, and its retardation in coming to rest, should each occupy about 15% of the stroking time of the cylinder;
- thus, the fastest required response of the system is as illustrated on Fig. 8.7b;
- for economy, the smallest pump and cylinder should be sought;
- standard hydraulic components are to be used;
- maximum system pressure allowed is 13.8 MPa.

Fig. 8.7c shows the main components of the hydraulic system proposed. It was decided to use an open-centre control valve. The high gain between valve motion and control flow in the valve's central position allows good control of load speed without the need for constant pressure supply. The counterbalance valve in the lower control line is needed to hold the load against gravity. The pressure relief valve in the upper control line is used to limit cylinder pressure (and hence force and acceleration) when the load is to be driven downwards, gravity providing much of the force needed for the downward stroke.

Conventional static design criteria were used

- to design the beam —— structural stiffness was the main consideration;
- to select the pump and cylinder —— this was associated with the selection of beam lever ratio, with requirement to use system pressure maximum of 13.8 MPa, and with the desire to use the smallest pump and cylinder which would provide the speed of load motion required (overall 0.6 m/s);
- to select the valves.

The results were

. beam length (from pivot to load centre)	1.52 m
. beam lever ratio (ratio load to actuator displacement)	5.1
. effective total inertia of driven system (load 125 kg plus actuator moving parts and beam inertias referred	141 kg

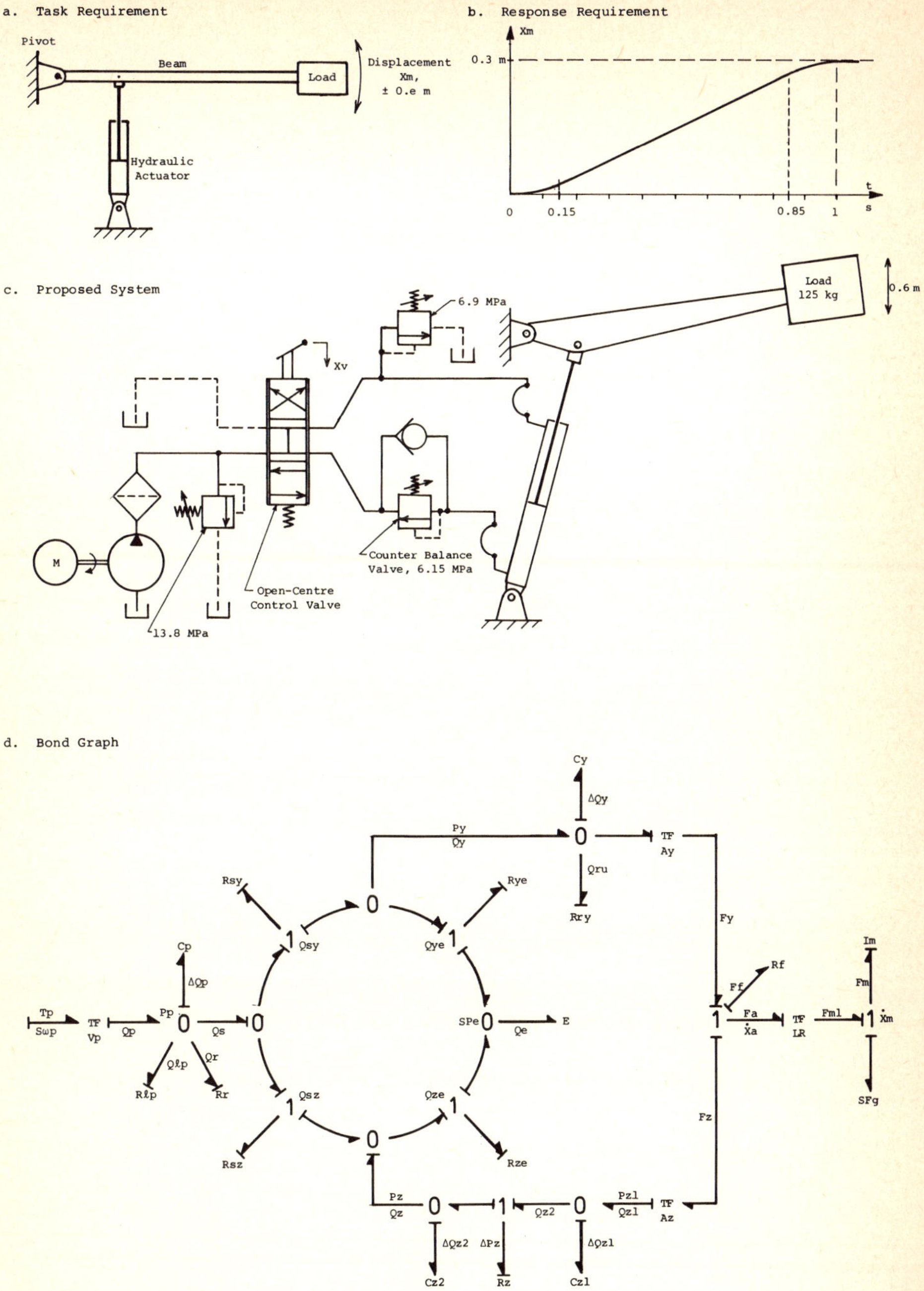

FIG. 8.7 LOAD LIFTING SYSTEM (EXAMPLE 8.7, TABLES 8.6 AND 8.7)

to the load, 16 kg)

. main system pressure relief valve setting (as specified)	13.8 MPa
. counterbalance valve pressure setting (required to support load)	6.15 MPa
. downstroke relief valve setting (required to limit speed of downstroke)	6.9 MPa
. hydraulic actuator —— bore	3.8 cm
—— stroke	0.33 m
—— rod diameter	2.54 cm
. hydraulic pump, fixed displacement	9.8×10^{-6} m^3/rev
. control valve	12 ℓ/min, 14 MPa

This brought the system design to the stage where a reasonably accurate assessment of its dynamic response was needed to see if it satisfied the speed of response, and the acceleration limiting specifications.

8.6.2 Development of Bond Graph

Fig. 8.7d shows a bond graph for the system developed on the bases that

- pump drive speed is constant
- positive valve action Xv causes lowering of the load (pos. Xm)
- the resistance of the short hydraulic lines have a negligible effect on load position response
- the capacitances of the lines can be lumped with Cp for the pump-valve lines, and with Cy and Cz for the valve-actuator lines
- the resistance of the filter is negligible, and its capacitance can be lumped into Cp
- the beam can be assumed to supply a constant lever ratio between actuator displacement and vertical displacement of the load Xm .

Component bond graph structures utilized are similar to those developed in Chapter 7 and used in previous examples. The driven load is modified to allow for the lever ratio effect. Power flow directions are shown as positive to the Y side and from the Z side of the actuator.

8.7.3 The Equations and Coefficients

The main purpose of the model is to allow prediction of the time response motion of the driven load following command inputs applied to the control valve. The bond graph of Fig. 8.7d allows development of equations for modelling the system for both lifting and lowering of the load. As seen in Section 7.9.5, many equations are required for comprehensive modelling of the open-centre valve. In the present system, lowering of the load is greatly assisted by gravity with little hydraulic power requirement, the main need being to control the speed of descent. Thus lifting of the load with hydraulic power is the main dynamic aspect. The equations of the model are given in Table 8.6. They are written for lifting only.

The coefficients of the model are shown in Table 8.7.

8.7.4 Simulation

Fig. 8.8 shows the input and the responses of some of the system variables when the model was simulated with MIMIC on a CDC Cyber 6400 computer. Load displacement Xm(t) is directly related to piston displacement Xa(t) by the beam lever ratio. Corresponding results obtained from the hardware system, which was assembled *subsequent* to the simulation, are included on Fig. 8.8. Considering the ambiguity of some model parameter coefficients, the correspondence is good. It is a simple matter to improve the correspondence between simulated and actual system performance by adjustment of some of the model's parameters.

Comparison between Figs. 8.7b and 8.8b shows that the specified response is approached but not achieved. The main discrepancy is the hump near $t = 0.5$ s . Some parameter adjustment, possibly using optimization procedures of the type to be discussed in Chapter 9, is called for. With the simulation working, it is a simple matter to study the effect on response of adjusting a parameter of the model.

8.7.5 Conclusion

This example is developed more fully in Refs. 8.4 and 8.5.

8.8 A HIGHLY DYNAMIC ELECTROHYDRAULIC CONTROL SYSTEM

8.8.1 Introduction

Hydraulic control systems are commonly used where the precise positioning of large resisting loads is required. High pressure hydraulics has little logic capability, being a power transfer technology. Where logic or feedback is required the hydraulic power system usually has to be interfaced with manual, mechanical, pneumatic, electronic, or electrical control logic. Electrohydraulic servo-systems,

TABLE 8.6 Model for Example 8.7

Component, Effect, Comment	*Equation*	
Hydraulic Supply		
Speed source	$S\omega p = \text{constant}$	8.7.1
Pump pressure, Cp effect	$Pp = 1/Cp \cdot \int \Delta Qp\, dt + Pp(o)$	8.7.2
Internal leakage, Rℓp effect	$Q\ell p = K\ell p \cdot Pp$	8.7.3
Relief valve, Rr effect	$Qr = 0$ if $Pp \leq Pset$ $ = Kr \cdot (Pp - Pset)$ if $Pp > Pset$	8.7.4
TF effect	$Qp = S\omega p \cdot Vp$	8.7.5
	$Tp = Pp \cdot Vp$	8.7.6
Pump 0-junction	$\Delta Qp = Qp - Q\ell p - Qr - Qs$	8.7.7
Control Valve		
Xv is spool displacement from centre.	For all Xv , 0 to ± Xv max	
	$Qs = Qsy + Qsz$	8.7.8
Xv max is that spool displacement which just fully opens valve. It is assumed that Xv cannot exceed Xv max .	$Qe = Qye + Qze$	8.7.9
	$Qy = Qsy - Qye$	8.7.10
	$Qz = Qsz - Qze$	8.7.11
	$Qsy = Ksy \cdot \{X(o) + Xv\} \cdot \Delta Psy^{\frac{1}{2}}$	8.7.12
X(o) is underlap.	$Qye = Kye \cdot \{X(o) - Xv\} \cdot \Delta Pye^{\frac{1}{2}}$	8.7.13
It is allowed that K values may be different.	$Qsz = Ksz \cdot \{X(o) - Xv\} \cdot \Delta Psz^{\frac{1}{2}}$	8.7.14
Note that there are	$Qze = Kze \cdot \{X(o) + Xv\} \cdot \Delta Pze^{\frac{1}{2}}$	8.7.15
. 4 0-junction equations	$\Delta Psy = Ps - Py$	8.7.16
. 4 R equations	$\Delta Pye = Py - Pe$	8.7.17
. 4 1-junction equations	$\Delta Psz = Ps - Pz$	8.7.18
in the basic model, one for each physical effect on the bond graph.	$\Delta Pze = Pz - Pe$	8.7.19
Assumes openings equal for all four orifices when spool centred, i.e. equal underlaps.	For Xv = 0 only, in addition to equations 8.7.8 to 19 $\Delta Psy = \Delta Pye = \Delta Psz = \Delta Pze = (Pp - Pe)/2$	8.7.20
When $Xv \geq X(o)$, ports YE and SZ are closed.	For positive $Xv \geq X(o)$, in addition to equations 8.7.8 to 19 $Qye = Qsz = 0$	8.7.21
When $\lvert \text{negative } Xv \rvert \geq X(o)$, ports SY and ZE are closed.	For $\lvert \text{negative } Xv \rvert \geq X(o)$, in addition to equations 8.7.8 to 19 $Qsy = Qze = 0$	8.7.22
Y Side, Line and Relief Valve and Actuator		
Pressure development, Cy effect	$Py = 1/Cy \cdot \int \Delta Qy\, dt + Py(o)$	8.7.23
Relief valve, Rry effect	$Qry = 0$ if $Py \leq Py\ set$ $ = Kry \cdot (Py - Py\ set)$ if $Py > Py\ set$	8.7.24
0-junction	$\Delta Qy = Qy - Qay - Qry$	8.7.25

TABLE 8.6 (contd.)

Component, Effect, Comment	*Equation*	
TF (Ay) effect	$Qay = \dot{X}a \cdot Ay$	8.7.26
	$Fy = Py \cdot Ay$	8.7.27
Z Side, Line and Actuator and Counterbalance Valve		
	If Xv is positive	
Cz1 is capacitance between actuator and counterbalance valve.	$Pz1 = 1/Cz1 \cdot \int \Delta Qz1 \, dt + Pz1(o)$	8.7.28
Pz1(o) is the set value of the counterbalance valve.		
TF (Az) effect	$Fz = Pz1 \cdot Az$	8.7.29
	$Qz1 = \dot{X}a \cdot Az$	8.7.30
Cz1 0-junction	$\Delta Qz1 = Qz1 - Qz2$	8.7.31
Counterbalance valve action, Rz effect	$Qz2 = 0$ if $Pz1 < Pz1(o)$ $= Qz1 - \Delta Qz1$ if $Pz1 > Pz1(o)$	8.7.32
Rz 1-junction	$\Delta Pz = Pz1 - Pz$	8.7.33
Cz2 is capacitance between counterbalance valve and control valve.	$Pz = 1/Cz2 \cdot \int \Delta Qz2 \, dt + Pz(o)$	8.7.34
Cz2 0-junction	$\Delta Qz2 = Qz2 - Qz$	8.7.35
<u>Note 1</u>: When the counterbalance valve opens, the Rz effect becomes negligible, Pz1 and Pz identical, and Cz1 and Cz2 can be lumped as Cz .		
<u>Note 2</u>: Equations 8.7.28 to 35 hold also for Xv negative so long as the counterbalance valve is seen to be either full open or full shut depending on the value of Pz1 and when open has negligible resistance to flow.		
Beam and Load Systems		
Net actuator force	$Fa = Fy - Fz - Ff$	8.7.36
Friction, Rf effect	$Ff = \text{sign } \dot{X}a \cdot F\mu + Kf \cdot \dot{X}a$	8.7.37
Load motion, Im effect	$\dot{X}m = 1/Im \cdot \int Fm \, dt + \dot{X}m(o)$	8.7.38
Net force	$Fm = Fm1 + SFg$	8.7.39
Gravitational force	$SFg = \text{constant}$	8.7.40
Beam power transformation TF(LR)	$\dot{X}a = \dot{X}m/LR$	8.7.41
	$Fm1 = Fa/LR$	8.7.42

TABLE 8.6 (contd.)

Component, Effect, Comment	*Equation*	
Auxiliary Equations		
	$Xm = \int \dot{X}m\, dt + Xm(o)$	8.7.43
Vy(o) is Y side volume under compression. By is Y side bulk modulus.	$Cy = (Vy(o) + Ay \cdot Xa)/By$	8.7.44
Vzl(o) is volume under compression between actuator and counter-balance valve. Bzl is related bulk modulus.	$Czl = (Vzl(o) + Az \cdot Xa)/Bzl$	8.7.45

TABLE 8.7 Coefficients for Example 8.7

Power Supply	Sωp	= 151 r/s
	Vp	= 1.56 E-6 m^3/r
	Kℓp	= 2 E-12 $(m^3/s)/Pa$
	Kr	= 2.5 E-9 $(m^3/s)/Pa$
	Pset	= 13.8 E6 Pa
	Pp(o)	= 0.5 E6 Pa
	Cp	= 2.91 E-12 m^3/Pa
Control Valve	Ksy	= Ksz = 10.5 E-6 $m^2/(Pa^{1/2} \cdot s)$
	Kye	= Kze = 13.7 E-6 $m^2/(Pa^{1/2} \cdot s)$
	Pe	= 0
	X(o)	= 0.019 m
Actuator, Y side	Py(o)	= 0.5 E6 Pa
	Kry	= 2.5 E-9 $(m^3/s)/Pa$
	Py set	= 13.8 E6 Pa
	Ay	= 0.63 E-3 m^3
Actuator, Z side	Pzl(o)	= 0.5 E6 Pa
	Az	= 1.14 E-3 m^3
	Cz2	= 0.2 E-12 m^3/Pa
	Pz(o)	= 0.5 E6 Pa
Beam and Load	Fμ	= 120 N
	Kf	= 1.57 E3 N · s/m
	Im	= 141 kg
	$\dot{X}m(o)$	= 0
	SFg	= 1.382 E3 N
	LR	= 5.1
Auxiliary Equations	Xm(o)	= 0
	Vy(o)	= 0.25 E-3 m^3
	By	= 380 E6 Pa
	Vzl(o)	= 0.22 E-3 m^3
	Bzl	= 535 E6 Pa

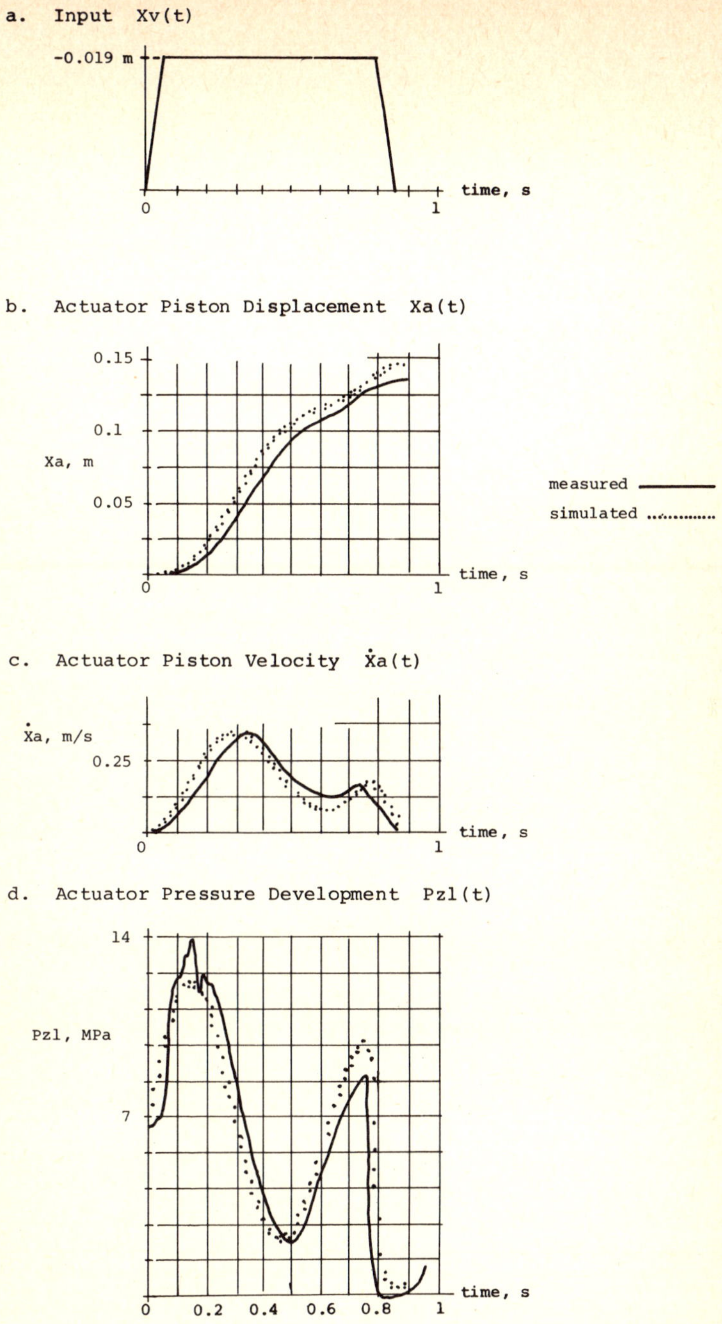

FIG. 8.8 RESPONSE OF LOAD LIFTING SYSTEM

long used in military and space technology, are now widely utilized in industrial applications such as machine tool control, flight simulators, structure fatigue testing rigs, computer-aided manufacture, and industrial robots. With increasing use of microprocessors and computer control it can be expected that electrohydraulic servovalves will be used increasingly in industrial and manufacturing automation.

Electrohydraulic drives are often used where the dynamic response requirements of the driven load(s) in terms of speed, cycling, accuracy, and stability, are particularly severe. One such application, the subject of the present Example, is the control and drive system associated with a 28 tonne seismic block facility. The intention is that the electrohydraulic system shakes, at controlled force and displacement amplitudes and frequencies, a structure or machine undergoing vibrations analysis. The present system is intended to be useful to 300 Hz. Fig. 8.9a illustrates the shaker system layout and Fig. 8.9b the essentials of the hydraulic system. The hydraulic power supply centres on a 30 kW Denison piston pump operating at up to 21 MPa set pressure and up to 60 ℓ/min flowrate.

A 30 kW induction motor drives the axial piston pump at 1000 rpm. The main pressure control valve is set at between 2.76 and 20.7 MPa, depending on requirements, to provide constant pressure at that value to the servovalve. The return line pressure control valve is set at between 2.1 and 3.45 MPa, depending on the system performance requirements, to prevent the possibility of cavitation with consequent dynamic performance uncertainty. Both pressure control valves are pilot operated by remote control from the servo-controller. The electrohydraulic servovalve receives its position feedback modulated input signal from the servo-controller. The water-cooled oil cooler enables oil temperature to be controlled, thus controlling oil viscosity, an important factor in high performance hydraulics. The long hydraulic hoses provide flexibility so that the actuator can be taken to any desired position on or around the seismic block (approximate dimensions 3 m × 4.5 m area × 0.65 m depth × 28 tonnes mass). The accumulator package provides the smoothing of pressure transients and the capacitive effects needed for smooth load position control. The electrohydraulic system has various override, safety, and functional features not included on Fig. 8.9 as they have little effect on dynamic performance under normal system operation.

8.8.2 The Bond Graph

In order to better understand the dynamic nature and performance of the system beyond the rather meagre information supplied by the manufacturer, a simulation of the shaker system was required. Because the main interest lay in employing the upper limits of dynamic response, the model for simulation needed to be detailed, accurate, and versatile.

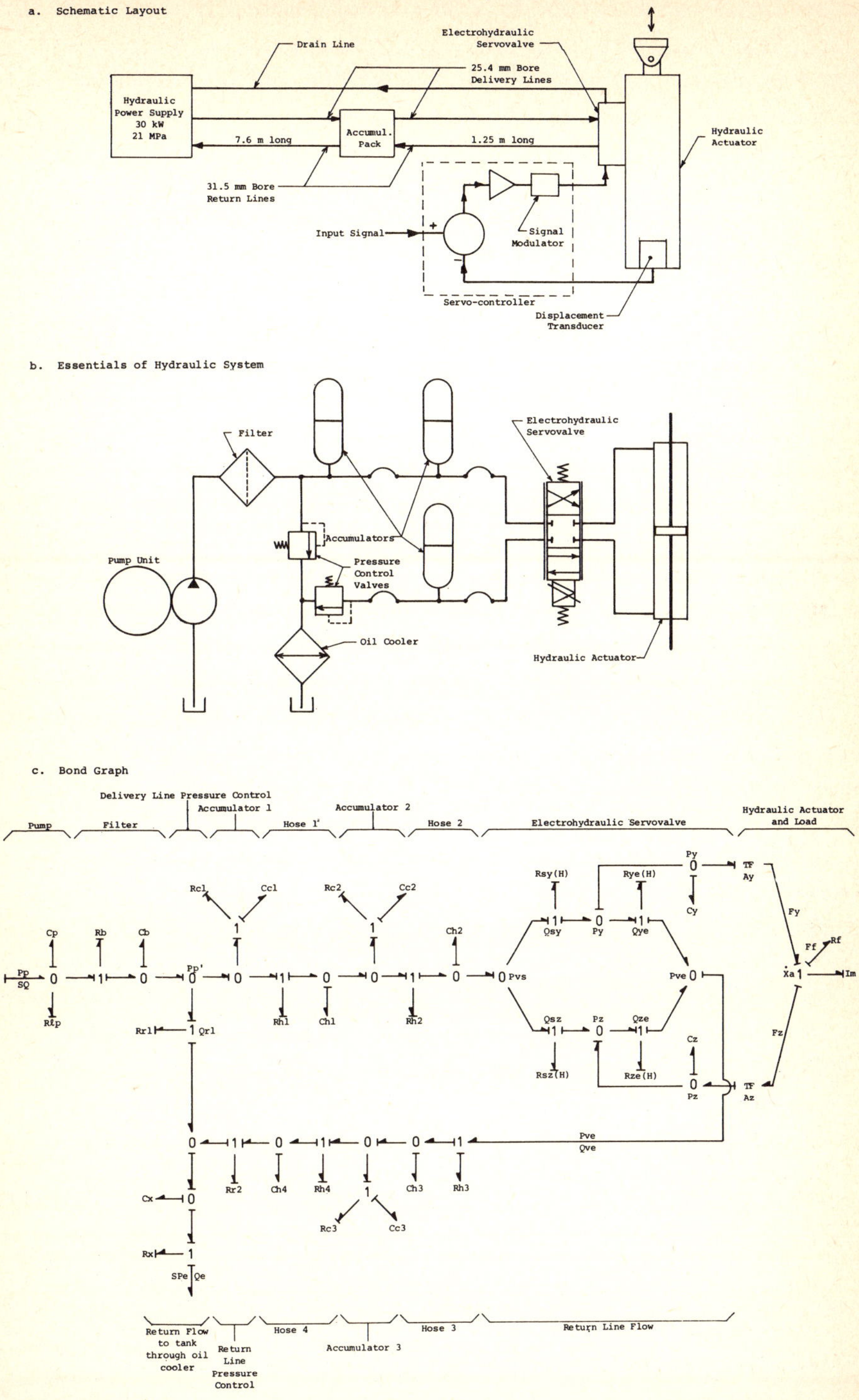

FIG. 8.9 AN ELECTROHYDRAULIC SYSTEM (EXAMPLE 8.7)

Fig. 8.9c shows the power bond graph structure developed for the system, the various hardware components being indicated adjacent to the appropriate sections of the bond graph. Some of the R and C effects included will not be simple coefficients. Discontinuous and other non-linearities are present.

The pump is represented as a constant flow source SQ (constant angular speed and constant swept volume rate). Internal pump leakage is accounted for as Rℓp , and the resistive and capacitive effect of the pump's discharge side is recognized as the combination Rb and Cb . Next is the main system pressure regulator, included as effect Rr1 . This regulator sets the system pressure. Proceeding right along the main power flow line, the hydraulic accumulators and the hydraulic hoses are allowed for as appropriate pairs of R and C effects.

The closed-centre electrohydraulic servovalve is of single-stage design. Its internal leakage, flow forces, friction, spool mass, and torque motor characteristics have all been neglected in the proposed bond graph. This is justified by the present interest being centred on the performance of the actuator and driven load and not on the servovalve itself. The bond graph for the servovalve is identical to those used previously for closed-centre valves (Figs. 7.15, 8.1, 8.2). It is necessary only to realize that Rsy , Rsz , Rye , and Rze are automatically and continuously adjusted by the electrical signal from the servo-controller. This is indicated on the bond graph by Rsy(H) etc. This input plus feedback situation could have been indicated via signal (non-power) bonds on the bond graph leading to each of Rsy , Rsz , Rye , and Rze , as illustrated on Fig. 8.10. Servo-controller control of pressures Pr1 and Pr2 could similarly be indicated via signal bonds from the servo-controller to Rr1 and Rr2 respectively. However these signal bonds clutter the bond graph. To a person familiar with the technology, as any designer or analyst should be, and appreciative of the relationships between Figs. 8.9a, b, and c, the modulation of the R effects noted should be obvious without the inclusion of the signal bonds, and even without use of the R(H) or R (input and feedback effects) notation.

The actuator and driven load sections of the bond graph are similar to those used in a number of earlier examples. The return line bond graph includes recognition of its hoses, its accumulator, and its pressure control valve (Rr2) . The return line from the EHSV joins with the discharge from the supply line pressure control (Rr1) , and the total discharge passes through the oil cooler to return to tank. The capacitance and resistance effects of the oil cooler and its associated lines are included on the bond graph as Cx and Rx .

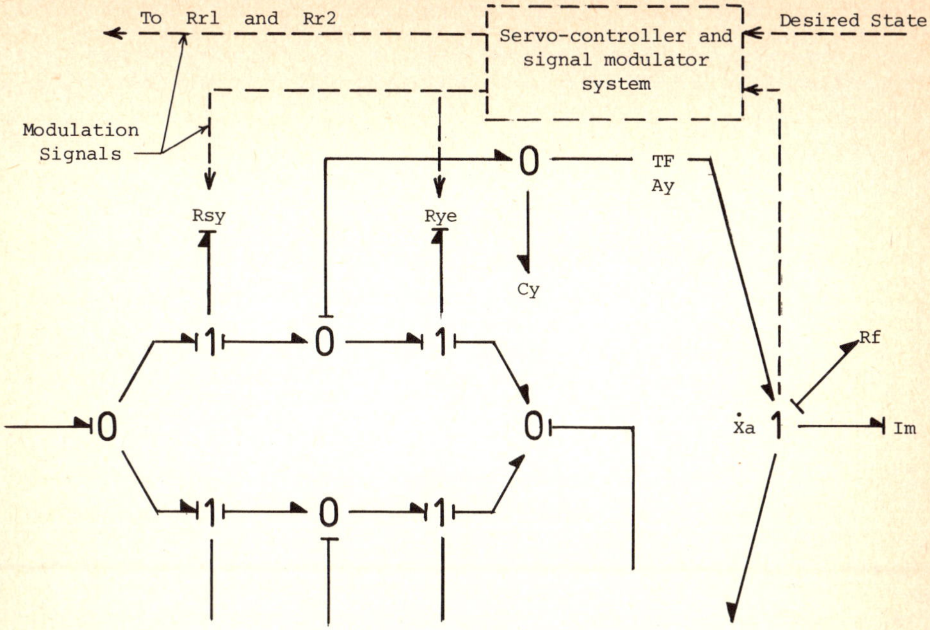

FIG. 8.10 DIRECT INCLUSION ON BOND GRAPH OF MODULATING SIGNALS

8.8.3 Conclusion

Further explanation of bond graph development, the equations derived using the bond graph, and results of a simulation of the system are given in Refs. 8.6 and 8.7. There are 77 equations in the model, most of forms already used in this Text. It should not be thought that the system bond graph of Fig. 8.9c came easily, without effort. To form and link the component bond graph structures took considerable effort in satisfying physical reality, adequate completeness, and causality. However, development of the bond graph provided the key to orderly and progressive model development in a situation which would have been daunting if approached by conventional modelling procedures.

8.9 CONCLUSION

A number of examples has been given of the use of bond graphs to provide a link between a hydraulic control system, real or proposed, and its simulation for analysis and synthesis purposes.

8.9 REFERENCES

8.1 Winton, R.D., An Investigation into the Mathematical Modelling of Hydraulic Control Systems, Using Power Bond Graph, Project Report, Dept. Mech. Eng., Monash University, Australia, 1973.

8.2 Dransfield, P. and Winton, R.D., Applying Power Bond Graphs to Hydraulic Control Systems, Proc. III Congresso Brasiliero de Engenharia Mecanica, Rio de Janeiro, 1975.

8.3 Stecki, J.S. and Cuthbertson, S.C., already cited as Ref. 4.3.

8.4 Barnard, B.W., The Dynamic Response of a Hydraulic System, M.Eng.Sc. Thesis, Monash University, Australia, 1973.

8.5 Barnard, B.W. and Dransfield, P., already cited as Ref. 7.4.

8.6 Teo, M.K., Development of a Power Bond Graph Model of an Electrohydraulic Vibrator System, Project Report, Dept. Mech. Eng., Monash University, Australia, 1974.

8.7 Dransfield, P. and Teo, M.K., Using Bond Graphs in Simulating an Electrohydraulic System, J. Franklin Inst., V.303, N.3, Pergamon, 1979.

CHAPTER 9

OPTIMISING SYSTEM RESPONSE

9.1 INTRODUCTION

The ethos of this Text is that ensuring good dynamic response should be an integral part of the design of a proposed hydraulic control system. Further, that the securing of this aim is best effected via mathematical modelling of the proposed system and digital simulation of the model. Further still that bond graph techniques provide a dynamic modelling procedure particularly suited to the modelling of hydraulic control systems.

To here the Text has concentrated on developing the bond graph procedures needed to model hydraulic control systems. It has been implicit that such models are best simulated by using digital simulation.

Having obtained a predicted response questions arise as to its quality. As illustrated in Chapter 1, Fig. 1.3, a predicted response might be satisfactory, promising, or hopeless. If promising, the designer-analyst should be able to develop his proposal towards the satisfactory state. If hopeless, he may have to consider a more drastic re-design to at least get model performance to a promising state.

In the present Text, optimization of response is regarded as the process of causing a promising response to converge on a desired response by adjustment of some model parameters. The model is optimized if its response shows the minimum deviation from the desired response possible under the variations of parameters allowed or practicable. Thus if a model is optimized it is not implied that its response has converged fully on the desired response. Rather, that it has converged on it as closely as is practicable. A system model and the allowed range of its parameters may be incapable of producing the precise required response. An optimum response is not necessarily acceptable.

As seen in the examples of Chapter 8, a system's model will have many coefficients. To retain satisfactory power and motion characteristics most of the coefficients will not be able to be varied much without basic re-design of the system. However, there are usually a few coefficients for which the designer-analyst has a reasonable choice of values. The idea becomes to seek that combination of values for these parameters which yields optimum response.

In the simplest case only one parameter is varied. A set of discrete solutions of the system model, with the chosen parameter varied over its allowed range, quickly leads to the optimum value of the parameter; that is, that value which produces the desired system response, or the possible response closest to the desired response. In the more likely case where two or more parameters can be varied simultaneously, the number of possible combinations of values of the chosen parameters is so formidable that computer techniques are required for the search for the optimal combination of parameter values —— again, that combination which provides response closest to that desired.

This Chapter describes a response optimization scheme that has been developed and used with hydraulic control system response (Refs. 9.1, 9.2). There is no attempt to generally survey control system response optimization.

9.2 THE REQUIREMENTS

Requirements for the performance of analytical response optimization of the kind envisaged are:

1. Specification of a realistic response-inducing input.

2. Specification of the response desired.

3. Development of a mathematical model which describes the system's dynamic performance and relates response to input.

4. Means to readily effect solutions of the model to provide predictions of system response.

5. Means to compare desired $(X1(t))$ and predicted $(X2(t))$ responses leading to description of error as

$$E(t) = X1(t) - X2(t)$$

6. A criterion for judging the overall value or significance of the error function E(t) such that minimization of the criterion minimizes the effect of the error and maximizes the quality of the response relative to the desired response.

7. A multiple-parameter search procedure capable of exposing that combination of values for the parameters chosen for optimization purposes which provides optimum response.

8. Means whereby the designer-analyst can quickly and reliably effect Step 3. It has been proposed that power bond graph techniques provide this.

9. Means whereby Steps 4 to 7 can be managed by the designer-analyst with a high degree of automation. The digital computer provides this facility.

9.3 ERROR CRITERIA

Fig. 9.1 illustrates a desired response X1(t) and a predicted response X2(t) . Error E(t) is represented by the shaded area between the two curves. If the shaded area can be eliminated, we have perfect optimization. If it can be minimized, by using allowed variation of selected parameters of the model to adjust X2(t) , the response has been optimized. Thus, the most obvious error criterion is the minimization of the total summation of error, i.e. minimization of the shaded area between X1(t) and X2(t) . The error criterion is called *minimization of absolute value of error*. The shaded area can be described

$$IAE = \int_0^t |E| \, dt$$

where IAE denotes Integral of Absolute value of Error

$|E|$ denotes absolute value of error

t is the time over which the comparison of X1(t) and X2(t) is made, measured from the time (t = 0) when the input is applied.

Minimization of the *error function* IAE becomes *the absolute error criterion*.

Many variations in the basic error function have been developed to bias or weight the error minimization procedure to be sensitive to particular characteristics of a system's response potential. Table 9.1 shows a number of error functions.

La Brooy [9.1] considered most of these in relation to responses typically required of hydraulic control systems. It was confirmed or concluded that

- IAE is less sensitive to large error and more sensitive to small error relative to ISE and is therefore likely to produce an optimum with larger

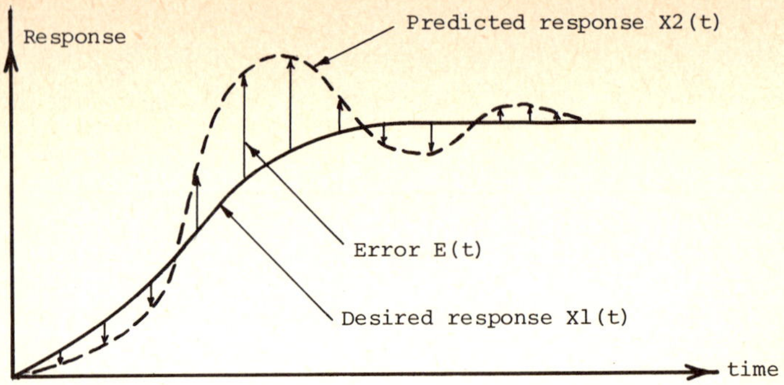

FIG. 9.1 ILLUSTRATING DYNAMIC ERROR

TABLE 9.1 Error Functions

Error Function Designation	*Definition*
ISE	$\int_o^{ts} E \cdot E \, dt$
ItSE	$\int_o^{ts} t \cdot E \cdot E \, dt$
IStSE	$\int_o^{ts} t \cdot t \cdot E \cdot E \, dt$
IAE	$\int_o^{ts} \lvert E \rvert \, dt$
ItAE	$\int_o^{ts} t \cdot \lvert E \rvert \, dt$
IXSE	$\int_o^{ts} e^{2Kt} \cdot E \cdot E \, dt$
IMXSE	$\int_o^{ts} \left[1 - e^{-2Kt}\right] \cdot E \cdot E \, dt$

I denotes 'integral of'
S denotes 'square of'
E denotes error
t denotes time
A denotes 'absolute value of'
X denotes 'exponential weighting of'
M denotes 'modified value of'
ts is simulation time (> settling time of model response)
e is the exponential
K is the arbitrary constant used with IXSE and IMXSE

rise time but less overshoot and oscillation,

- ISE is heavily influenced by the initial stages of response, when error is largest, but is insensitive to lingering oscillations of small amplitude, and is likely to produce an optimum response with light damping, fast rise time, considerable overshoot, and considerable oscillation,

- time weighting (ItSE , ItAE , IStSE) reduces the contribution of initial error and increases the contribution of prolonged error. This is likely to encourage shorter settling time. However, it may inhibit fast initial response,

- the exponentially time-weighted functions (IXSE , IMXSE) , with appropriate choice of K , allow a wide range of error function characteristics to be obtained. These functions weight prolonged error exponentially and are likely to yield optimum responses more heavily damped than do the other functions. Judicious selection of K can begin the exponential weighting at any point in the time frame of the response.

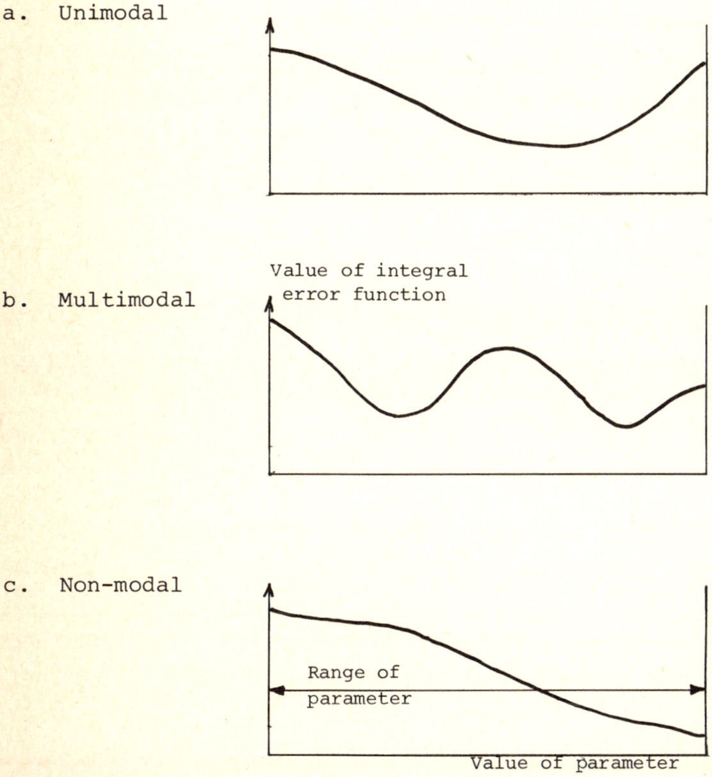

FIG. 9.2 SOME ERROR FUNCTION SHAPES

Hydraulic control systems are stiff and are usually heavily loaded. They are expected to inhibit fast response to step inputs for example, possibly with some overshoot and oscillation. It appears that ItAE is suitable as an error function in these circumstances, and it is adopted for the example of this Chapter. However, this should not be interpreted as endorsement of ItAE to the general exclusion of other error functions.

9.4 SEARCH PROCEDURES

9.4.1 Introduction

The form of the value of an error function as optimization parameters are changed over their allowed ranges can be

- unimodal, possessing a minimum as illustrated for single parameter optimization on Fig. 9.2a,

- multimodal, possessing several minima, Fig. 9.2b,

- non-modal, possessing a minimum at one of the parameter boundaries, Fig. 9.2c.

The unimodal case is the most straightforward. The multimodal situation can cause problems, as a search procedure might direct itself towards a local rather than the global minima. The non-modal case appears to be unambiguous; however it can cause computational problems as the search heads to the hard boundary at which the minima is located. Prior to performing a detailed error function analysis, a designer-analyst would rarely know which of the three situations of Fig. 9.2 he has. A good search procedure has to be able to seek a global rather than a local minimum, and has to be capable of negotiating hard boundaries. Constrained limited multi-parameter optimization is required.

9.4.2 Single Parameter Optimization

Young [9.3] considered that the Golden Section Search Procedure described by Meichke [9.4] was efficient and effective for incorporation into the digital simulation procedure MIMIC to provide an automated single-parameter optimization procedure suitable for the class of systems relevant to the present Text. Initial error function evaluations at fixed interval over the range of the chosen optimization parameter can reveal the region in which the true minimum exists. Multimodal minima is unlikely to be a problem with hydraulic control systems as the range over which a parameter can be varied realistically is likely to be quite narrow. It is more likely that the error function will be unimodal or non-modal.

9.4.3 Multi-parameter Optimization

The error function shapes illustrated on Fig. 9.2 apply to single-parameter optimization. It can be seen that 1-parameter optimization required 2-dimensional error function representation. For 2-parameter optimization, the lines of Fig. 9.2 become surfaces; i.e. the error function for 2-parameter optimization requires 3-dimensional representation. In general, an error function is related to the n parameters of an n-parameter optimization scheme by a shape in (n + 1) space. When n is larger than 1 or 2, the shape becomes complex and abstract. Scanning such a space for an optimum combination of the n parameters is a substantial computational task. A manageable, efficient, and reliable search proceaure for handling constrained and at least 3-parameter optimization is required for hydraulic control system analysis.

The Golden Section Search Procedure for 1-dimensional optimization does not effectively extend to multi-parameter use [9.5]. More practical search techniques are based on the evolutionary operation method of Box [9.6].

The Simplex method [9.7] is one EVOP approach. A Simplex is defined generally as a shape formed by (n + 2) points distributed in (n + 1) dimensional space, n being the number of optimization parameters. Error function evaluations are made for each of the (n + 2) points of the Simplex. The point producing the worst evaluation, i.e. that point furthest from the point producing the smallest evaluation, is discarded and replaced by a point located in a more favourable position. This point is located using a set of mechanical rules intended to help the Simplex converge towards the point of minimum possible error evaluation. It has been reported [9.8] that Simplex is most satisfactory when the error function is linear and when no constraints are placed on the range of each optimization parameter. Also that Simplex is susceptible to collapse into a sub-space to yield a local rather than the global optimum. For these reasons, Simplex seems to be inadequate for optimization of the response of hydraulic control systems. It is recognized that Simplex techniques might have been improved to accommodate its earlier limitations.

The Complex search method proposed by Box [9.9] and improved by Guin [9.10] appears to be well suited to the present requirements. Complex is heuristic (self-learning), is capable of handling non-linear error functions with unequally constrained optimization parameters, can negotiate hard constraint boundaries, and can turn corners in the space being scanned. It is unlikely to collapse into a sub-space to yield a local minimum, as a single point existing in a "better" region (i.e. closer to the global minimum) can cause the complex to be re-directed towards this point.

9.4.4 Complex

A general complex of dimension sigma is defined as a figure formed by sigma points distributed in the (n + 1) space of the n-parameter optimization scheme. Sigma must be $\geqslant (n + 1)$, and the optimization parameters explicit and independent of each other. The initial distribution of the sigma points of the complex is random except that at least one point must be within the optimization space specified, i.e. it must satisfy all constraints on the chosen optimization parameters. If other points lie outside the constraint boundaries, they are moved back to just inside by formal Complex procedures.

The random nature of the casting of the initial complex of points allows a probability that a point may be in the region of the global error function minimum which locates the optimal combination of optimization parameters. In this case, the Complex procedure will cause the complex to collapse towards the optimal point. If no point of the initial complex lies in the region of optimality, the heuristic rules governing the motion of the complex cause eventual scanning of the optimal region. Like Simplex, the rules of Complex generate new points to replace "worst" points with the objective of collapsing the complex towards the global optimum point. Like most search procedures, a wide choice of error function is allowed.

Fig. 9.3 illustrates, with a simple 2-parameter example, the nature of Complex. The ordinates show the boundaries of the two optimization parameters N1 and N2. The small-chain broken lines illustrate contours of constant values of the chosen error function as N1 and N2 are varied. Adopting sigma = 3 (i.e. at least n + 1), the three points of the complex are shown as N1, N2, N3, and their centroid is at $\bar{N}$. To minimize the value of error function, the complex must be directed from the 10 contour towards the 4 contour. This is effected by reflecting the worst point, N1, through and beyond the centroid to N1'. The next-stage complex becomes N2, N3, N1'. It can be seen that N3 would suffer the next reflection through the new centroid.

Fig. 9.4 shows the flow chart developed by La Brooy [9.1] for compiling the optimization algorithm COMPLEX which he used with MIMIC to provide an automated optimization procedure. The flow chart was modified from that of Box [9.9] for increased efficiency. A system model simulated in MIMIC calls on the sub-program COMPLEX which returns a parameter combination via multiple RETURN statements for MIMIC to effect a response solution. MIMIC then evaluates the error function for the particular parametric combination and recalls COMPLEX, passing the error function value and receiving another parameter combination in return. The sole purpose of the digital simulation procedure is to provide objective error function evaluations. COMPLEX implements the logic of Fig. 9.4 to enable the Complex procedure to function. It is a discrete Fortran optimization sub-program containing several parameters such

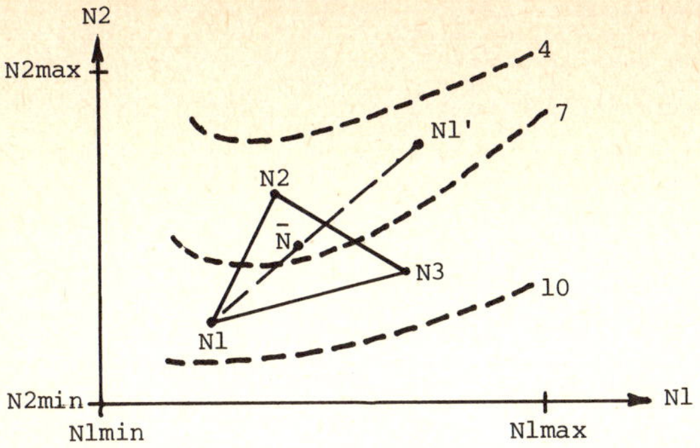

FIG. 9.3 REFLECTING A COMPLEX POINT

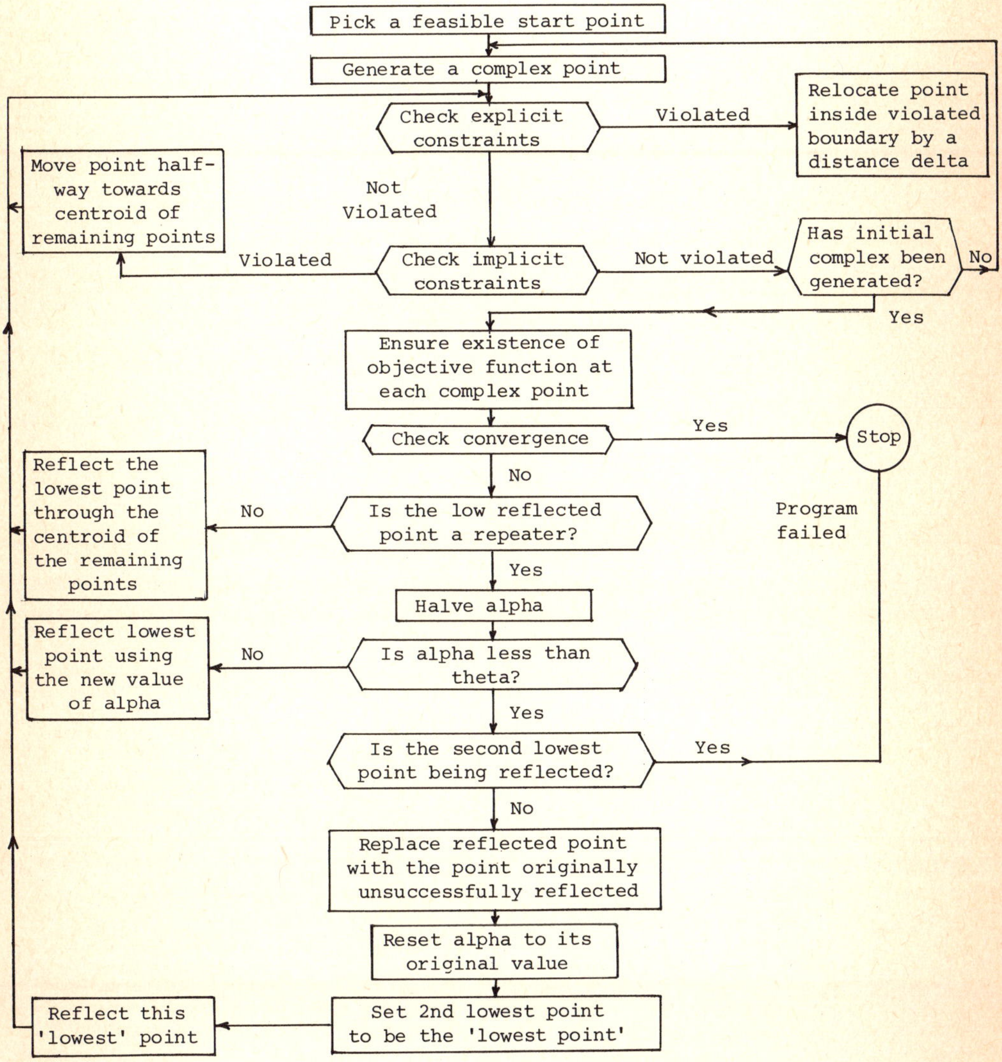

FIG. 9.4 FLOW CHART FOR COMPLEX

as ALPHA and BETA in Fig. 9.4 to which values appropriate to the particular problem can be ascribed. Usually one or two attempts at optimization are required before knowledge adequate for a successful optimization is gained. Used initially with MIMIC, COMPLEX was later combined with CSMP.

9.5 EXAMPLE

Fig. 9.5a and b illustrate a hydraulic servo-positioning system and its block diagram model [9.11]. Fig. 9.5c shows the load position response required for unit step input. The two servosystem coefficients K1 and K2 are available for system response optimization purposes in the ranges

$$0.01 < K1 < 0.05$$

$$0.3\ E\text{-}3 < K2 < 0.4\ E\text{-}3$$

The error function ItAE and a six point complex were chosen and incorporated in the COMPLEX-MIMIC package. One of the initial complex points was chosen to be K1 = 0.03 and K2 = 0.35 E-3 , mid-point in the available ranges. The remaining five points of the initial complex were randomly generated.

After 18 rejections of "worst" points of the complex —— i.e. 18 reflections and iterations —— error function values were within 30 E-6 of each other. The centroid of the final complex

$$K1 = 0.024$$

$$K2 = 0.3\ E\text{-}3$$

was accepted as optimization.

Fig. 9.5d shows the response of the optimized system. For comparison, Fig. 9.5e shows the responses for K1 and K2 at their bottom and top allowed values respectively.

Note that the optimum value for K2 is at one of the K2 boundaries. It is probable that an improved response would be achieved if the K2 boundary could be lowered. The same example with response criteria of shortest rise time, minimum overshoot, and minimum settling time following step input yielded the optimum condition K1 = 0.0182 , K2 = 0.323 E-3 , and the response shown of Fig. 9.5f. The responses for K1 and K2 at their bottom and top allowed values are included.

9.6 CONCLUSION

A certain combination of established steps and procedures which together form a

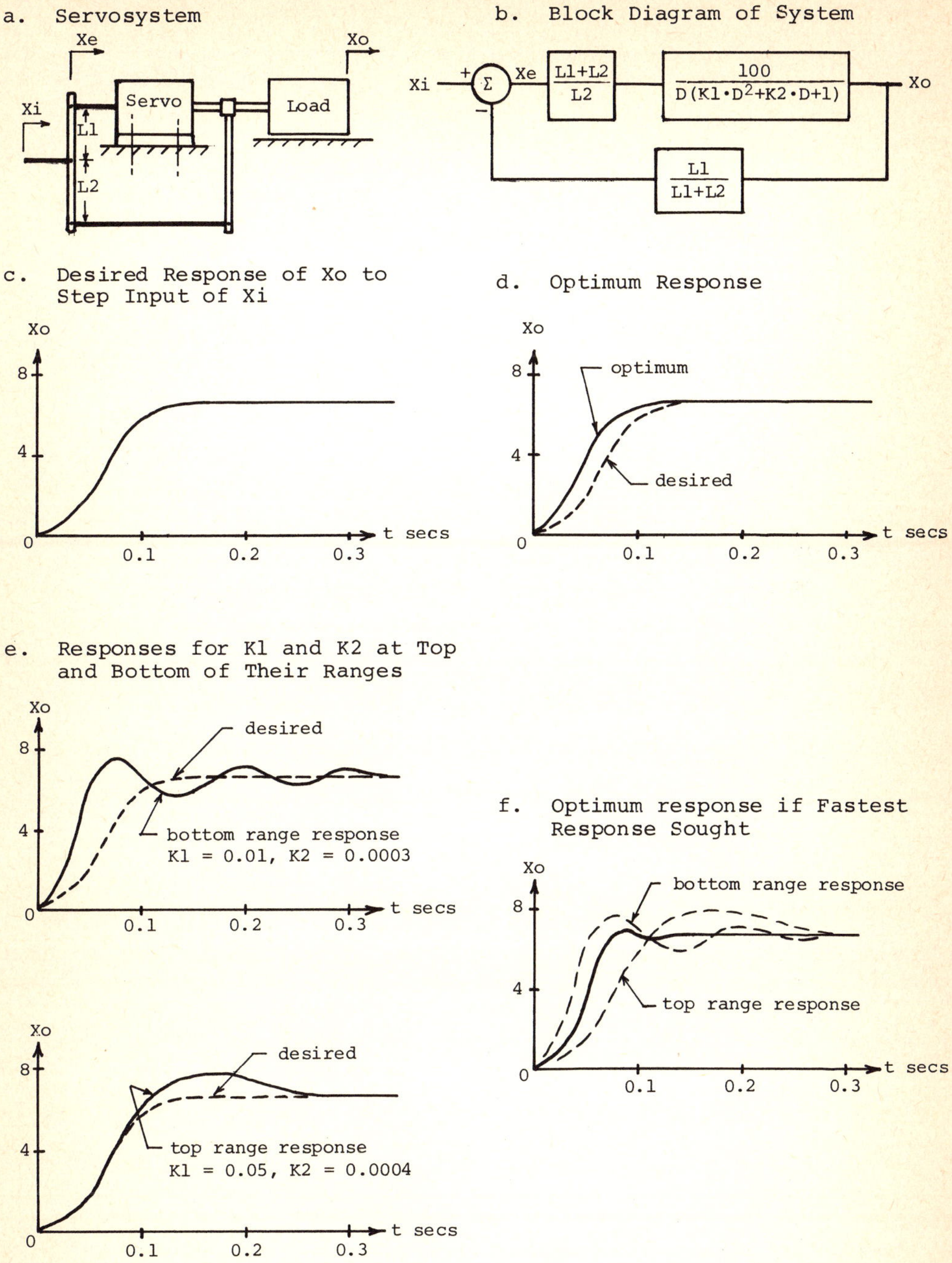

FIG. 9.5 EXAMPLE

scheme for optimizing the response of hydraulic control systems has been described and demonstrated. The component steps and procedures are:

- specification of a desired response to a specified input, in form, time, and magnitude
- mathematical representation of the system, and selection from within that model of several parameters for use in optimization
- selection of an appropriate integral error function whose minimization becomes the error criterion. ItAE is suggested for typical industrial hydraulic control system responses
- use of COMPLEX to search for the optimal combination of optimization parameters
- use of digital simulation to effect the repetitive model solutions required for estimations of error.

The example presented was solved on a COMPLEX-MIMIC or COMPLEX-CSMP package which integrated the steps and procedures required for optimization.

It is emphasized that optimization of the model's response is sought. The system will be optimized only if the model truly represents the system.

There are many variations of optimization procedures available in the literature.

9.7 REFERENCES

9.1 La Brooy, R., On the Optimisation of Hydraulic Control System Response, M.Eng.Sc. Thesis, Monash University, Australia, 1975.

9.2 La Brooy, R. and Dransfield, P., On the Time Response Optimisation of Hydraulic Control Systems, Proc. 4th Intl. Fluid Power Symp., BHRA, 1975.

9.3 Young, M.R., Digital Simulation of Hydraulic Control Systems, M.Eng.Sc. Thesis, Monash University, Australia, 1973.

9.4 Meichke, J., An Introduction to Computer-Aided Design, Prentice Hall, 1968.

9.5 Sugie, N., An Extension of Fibonaccian Searching to Multidimensional Cases, IEEE Trans. on Auto. Control, AC-9, 1964.

9.6 Box, G.E.P., Evolutionary Operation: A Method for Increasing Industrial Productivity, App. Stat. V6, N2, 1957.

9.7 Nelder, J.A. and Mead, R., A Simplex Method for Function Minimisation, Computer Jrl., V7, 1964-65.

9.8 Jacoby, S.L., Kowalik, J.S. and Pizzo, J.T., Iterative Methods of Non-linear Optimisation Problems, Prentice Hall, 1972.

9.9 Box, M.J., A New Method for Constained Optimisation and a Comparison with Other Methods, Computer Jrl., V8, 1965-66.

9.10 Guin, J.A., Modification of the Complex Method of Constrained Optimisation, Computer Jrl., V10, 1967-68.

9.11 Dransfield, P. and LaBrooy, R., Hydraulic Systems with Precision Reflexes, Machine Design, May 20, 1976. The example developed in this article came from Fluid Power Control, Blackburn, Reethof, and Shearer.

CHAPTER 10

PHENOMENA WHICH CAN AFFECT SYSTEM RESPONSE

10.1 INTRODUCTION

Phenomena which are inherent in working hydraulic control systems to the extent that they are always present have been included in the modelling procedures dealt with in this Text. Included in this *inherent phenomena* category are

- internal leakages through running clearances,
- viscous friction associated with components in relative motion,
- various R , C , and I effects,
- minimal coulomb friction.

There is another category of phenomena whose constituents may or may not be present and which if present will affect adversely the response of a hydraulic control system. These *parasitic phenomena* include

- excessive coulomb friction,
- cavitation,
- hydraulic backlash,
- valve flow forces,
- hydraulic lock,
- contaminated system fluid.

Some can be allowed for to some degree, if necessary, in system simulation models. It is better that components and systems be designed and operated so that parasitic phenomena do not appear. To achieve this requires an understanding of the phenomena and an appreciation of how each can be avoided, or its affects on system response minimized.

With the exception of coulomb friction, which has been considered already, each of the mentioned parasitic phenomena will be discussed briefly. References for further study are provided.

10.2 CAVITATION

10.2.1 General Discussion

A decision to use a hydraulic system for a particular control application is likely to be based on needs for

- fast and stable point-to-point positioning of large resisting loads, or speed control of large rotary loads,
- accurate and stiff location of positioned loads, or holding of rotary speeds,
- accurate control of force or torque.

To provide these characteristics, the fluid used in the system must retain a high bulk modulus as it passes around the operating system.

Quiescent hydraulic system fluids at normal temperatures and pressures have a high bulk modulus. That is, they are highly resistant to compression and have the potential to impose fast, stiff, and accurate drives on large resisting loads. Anything which reduces the high natural bulk modulus of a working hydraulic fluid, even temporarily, reduces the system's performance potential.

Quiescent hydraulic fluids contain gases. Typically, a hydraulic mineral oil at atmospheric temperature and pressure contains some 9% by volume of air. This air is not in the form of bubbles. It is dissolved in and is an integral part of the fluid. Oil in this normal state is said to be saturated. The amount of air which is absorbed and dissolved naturally in the oil is dependent on both pressure and temperature. At higher pressures more air will be absorbed and dissolved, and vice versa. Henry's law states that the solubility of a gas in a liquid is directly proportional to pressure. At higher temperatures less air remains dissolved in the oil, and vice versa.

If the pressure of a quiescent oil is changed slowly, air is absorbed and dissolved (for increasing pressure) or comes out of solution and is rejected (for decreasing pressure), as the natural saturation state appropriate to the particular pressure is sought. The mechanism of rejection of previously dissolved air as pressure is lowered is for bubbles of air to form and to float through the surface of the fluid.

This takes time, particularly with viscous fluids. Air is vastly more compressible than is oil, and consequently lowering of pressure has far more effect on the volume of the air component of the fluid than on the oil component.

The presence of dissolved air up to the natural saturation level has little effect on the bulk modulus of an oil. Thus, air which remains dissolved in the fluid will not affect the response of a hydraulic control system. However, the presence of bubbles of air in the oil will render the fluid more compressible, as any increase in pressure will quickly compress the bubbles of air. In these circumstances, system response can be adversely affected. Taken to the extreme, if a large quantity of air bubbles exists in the oil, the fluid suffers a drastic reduction in its bulk modulus.

The fluid in a working hydraulic control system experiences frequent, large, and sudden changes of pressure, in both directions. Thus, the saturation level of dissolved air will be changing rapidly and the actual level of dissolved air will be trying to adjust itself. This does not matter much, so far as system response is concerned, while pressures are increasing, as the fluid will be sub-saturated. However, when pressure is decreased rapidly, such as between a reservoir and a pump's suction port, or downstream of the orifice of a control valve, the fluid can be supersaturated and air coming out of solution as bubbles is a likely consequence. When the pressure drop is from one positive value to another positive value well above atmospheric pressure, this condition should not be serious. Nor should it seriously affect system performance. The bulk modulus will be transiently lowered but will remain high. If however the low pressure approaches atmospheric pressure or is below it, the volume of air coming out of solution can be large, the bulk modulus of the expanding fluid can approach zero, and system performance can be grossly degraded.

The formation of bubbles in the fluid due to air or gases coming out of solution due to reducing pressure is known as *gaseous cavitation*. In effect, the bubbles are cavities in the liquid. Such cavities are highly compressible and cause reduction of the fluid's bulk modulus. Once formed, gas cavities don't immediately dissolve if system pressure is again increased. It takes time for re-dissolving to occur. Thus a reduction in bulk modulus arising from gaseous cavitation can persist temporarily even if system pressure rises.

Quite apart from the natural gaseous cavitation described, a rapidly moving fluid undergoing pressure changes, directional changes, and temperature changes in the presence of air or other gases can entrain bubbles of the gas. This is likely to occur for example if return lines to a reservoir are not drowned, or if there are air leaks, no matter how minor, in uncharged pump suction lines. Entrained bubbles affect the fluid just as do bubbles of previously dissolved air. Thus the term *gaseous cavitation* can be regarded as including entrained undissolved gases.

Air entrained at low pressure can dissolve into the oil when pressure is increased, raising its air content level towards the saturation state. This is highly undesirable as the potential for gaseous cavitation, when pressure is inevitably lowered at a later part of the hydraulic circuit, is increased. The oil in a well-designed, well-operated, well-maintained hydraulic control system will tend to become sub-saturated. Any air entrainment will negate this desirable condition.

Like any liquids, hydraulic system fluids will boil (vapourize) if either the temperature is high enough or the pressure is low enough. Vapourization is characterized by the development in the liquid of vaporous bubbles which in effect are cavities in the liquid. Each liquid has a pressure, at which vapourization will develop. If pressure is reduced to or below this vapour pressure, vapourization will occur. A liquid's vapour pressure is temperature dependent, increasing as temperature is increased. Thus, fresh water boils at 100°C at sea-level atmospheric pressure 0.1 MPa. That is, at 100°C, its vapour pressure is 0.1 MPa. At temperatures below 100°C, the water's vapour pressure is below 0.1 MPa and if atmospheric pressure is maintained water remains in its liquid state. Water at 21°C will vapourize if its pressure is reduced to 0.0024 MPa, a strong vacuum state. Hydraulic fluids exhibit similar characteristics. Mineral oils have vapour pressure characteristics lower than those of water and are consequently less susceptible to cavitation. Water-based hydraulic fluids have vapour pressure characteristics more like water.

A fluid's susceptibility or its resistance to *vaporous cavitation* can be described by its cavitation number

$$Nc = \frac{P - Pv}{Y \cdot \rho \cdot \dot{X}^2}$$

where P is the static pressure present

Pv is the appropriate vapour pressure

ρ is the density of the fluid

$\dot{X}$ is the local velocity of the fluid

Y is a coefficient for the particular fluid.

Low Nc indicates a high tendency to cavitate and vice versa. Hence high Nc, which results from low Pv, low $\dot{X}$ and low ρ, is desirable.

. Summarizing, cavitation occurs if cavities develop in the working fluid. A cavity indicates the presence of a bubble of air, gas or vapour. Cavities can develop due to

- air or gas previously dissolved in the oil coming out of solution (gaseous cavitation)

- air or gas being entrained in the oil (gaseous cavitation)

- pressure of the oil falling below its vapour pressure, whereon the oil boils (vaporous cavitation).

In working hydraulic control systems, gaseous and vaporous cavitation can co-exist. Gaseous cavities are likely to form as soon as system pressure is reduced below the appropriate saturation value, which can be a quite high pressure. Vapour cavities will not form until pressure drops below the appropriate vapour pressure which is normally around or below atmospheric pressure.

- The presence of cavities in the working fluid reduces its bulk modulus. If the reduction is dramatic, system response will be degraded.

- The presence of dissolved air, and its coming and going as pressure changes, is a natural phenomena. Little can be done about it. Unless accompanied by vaporous cavitation, it is not usually of serious consequence. Entrained air is always undesirable and can seriously degrade system performance. Good reservoir design and good suction line design can minimize the presence and effect of entrainment. Vaporous cavitation can be avoided by keeping pressures at all parts of the system above the fluid's vapour pressure. If necessary, reservoirs can be pressurized and/or main pump suction lines charged.

- Cavitation not only degrades system response and performance. It can also damage the surfaces of components due to very high instantaneous pressures generated as cavities implode. It can also damage the system fluid by oxidation due to the very high local temperature generated as bubbles of air or gas or vapour are compressed when pressure increases. Cavitation also degrades a system's performance by generating offensive noise.

Refs. 10.1 to 10.3 provide more detailed discussions of cavitation and its relevance to hydraulic control systems.

10.2.2 Cavitation in Modelling and Simulation

Fully-developed vaporous cavitation can cause the bulk modulus of the system fluid in the region of cavitation to be zero. Under these conditions the pressure in the cavitating region is effectively zero and no power will be being transmitted from the region. The zero bulk modulus, zero pressure, zero power condition applies only to the region of cavitation and is usually highly transient.

A dynamic simulation indicates the presence of vaporous cavitation if it shows a pressure at or below atmospheric zero. Unless steps are taken to prevent it, a simulation can happily show and proceed with a negative pressure. As negative pressures (absolute) are unreal, a simulation becomes unrealistic if it allows the development of negative pressures. The pressure of a cavitating liquid remains effectively zero. However, it is not adequate in a simulation to simply hold a negative pressure development at zero until it again becomes positive. Though this appears to reflect the physical reality, it does not recognize the flow deficiencies associated with the computed negative pressures. That is, there is no recognition of cavity formation. This could lead to errors in load response predictions, though the errors would probably be transient in most situations.

McCloy [10.4] provides an approach for allowing for cavity formation in simulation models of hydraulic cylinders. He also suggests that the model complication required is rarely justified on the basis that its affects on cylinder response are minimal. Rogers [10.5] found it necessary to include modelling of cavity formation in simulation studies of hydraulic backlash, to be discussed in Section 10.3.

The concensus of opinion appears to be that while cavitation in hydraulic control systems remains an undesirable problem, its effects on load response are readily eliminated or minimized by established system design procedures, and that it is very rarely necessary to complicate dynamic simulation models by attempting to describe cavity formation so long as these procedures are utilized.

10.3 HYDRAULIC BACKLASH

One of the most common applications for hydraulic control systems requires the moving of an inertia load from one state (position or speed) to another in a specified range. There is usually the requirement or expectation that the driven load will be restrained accurately and stiffly in its new state prior to the next instruction, which might be milliseconds, seconds, minutes, or hours later. For the case of linear positioning, this objective is usually sought by using a hydraulic actuator whose output is rigidly connected to the driven load and which receives oil from a closed-centre or tandem-centre control valve. The intention is that when the control valve is centred, both hydraulic lines joining the valve to the actuator will be, and will remain, full of oil in a high bulk modulus state, thus stiffly locking the load in its required position. With rotary drives, the intention is that both the supply line and the discharge line of the rotary actuator be filled with oil of a high bulk modulus and which is flowing at a constant rate.

Unfortunately, these states are not always achieved, and the driven load can sometimes be easily moved about its desired state. This effect is called *hydraulic backlash*. Its basic characteristic is illustrated on Fig. 10.1. Ideally, any attempt to move the located load should immediately induce a high resisting force or torque from the hydraulic drive. If an attempt to displace the load meets zero or little resistance, at least over a band of displacement, hydraulic backlash exists. In the circumstances described, hydraulic backlash can exist only if the hydraulic fluid in one or both of the hydraulic lines of the actuator has a zero or very low bulk modulus. It is assumed that the hydraulic components have normal resistance to internal leakage and that the mechanical components of the system and load are of normal rigidity.

Not all hydraulic control systems are susceptible to the development of hydraulic backlash. The phenomenon will develop in susceptible systems only under some operating conditions, notably during rapid braking of inertia loads by valve or pump displacement control. It tends to be transitory in nature and usually decays in time after a load reaches equilibrium, though this time can be of significant duration.

Vaporous cavitation is pre-requisite for hydraulic backlash to develop, as a deficiency of liquid oil in an actuator's line(s) implies the presence of vapour. Hydraulic backlash is an effect or result possible if cavitation has occurred. Thus, anti-cavitation additions to a hydraulic circuit, such as the check and relief system illustrated on Fig. 10.2, act also to eliminate hydraulic backlash.

Rogers [10.5] made a substantial study of hydraulic backlash as it can develop during valve-controlled braking of a system of the kind illustrated on Fig. 10.2. With the cross-line relief system rendered ineffective, hydraulic backlash was quite substantial, up to 3 radians, and could persist for up to 20 minutes after the load was stopped. Rogers describes the mechanism of hydraulic backlash development. It is attributable to flow deficiencies in the actuator lines, relative to load motion, as the control valve is centred. Rogers' experiments and simulations show clearly the existence of both cavitation and hydraulic backlash as the rotating inertia load is slowed by braking. Rogers allowed the possibility of cavitation in his simulation model. Ref. 10.6 provides a description of some of this work.

In some circumstances, hydraulic backlash might provide the explanation for an observed response aberration or an aberration between an observed and a predicted response.

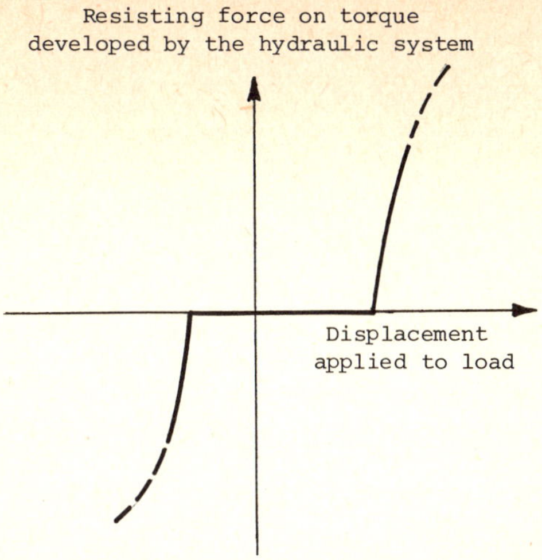

FIG. 10.1 HYDRAULIC BACKLASH CHARACTERISTIC

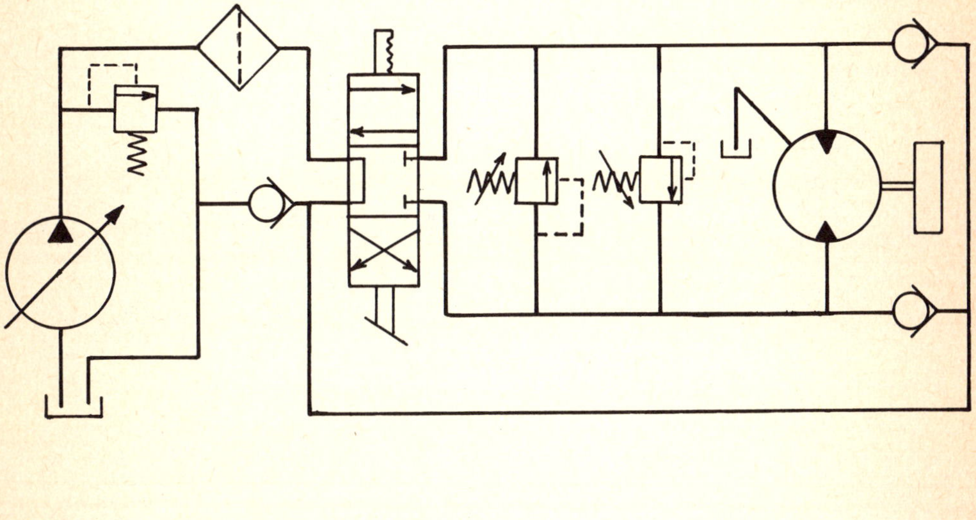

FIG. 10.2 ANTI-CAVITATION CIRCUIT ADDITIONS

10.4 FLOW FORCES IN VALVES

In this Text, the relationships required for including valve characteristics in the dynamic models of hydraulic control systems have been represented by R effects which directly relate output flowrates to input control signals. For example,

- the manually controlled 4-way valve has been represented by systems of equations centred around the relationship

$$Q = Kv \cdot Xv \cdot P^{\frac{1}{2}}$$

- pressure relief valves have been represented by

$$Qr = 0 \qquad \text{if} \quad P \leqslant Pset$$
$$\quad = Kr \cdot (P - Pset) \quad \text{if} \quad P > Pset$$

- electrohydraulic servovalves have been represented by systems of equations similar to those for manually-operated valves.

The thrust of the Text has been not the accurate simulation of individual hydraulic components, but rather the use of simplified representations of them which allow reasonable predictions of the major system response characteristics such as response of the driven load, and pressure developments in the system.

Valves experiencing flow develop force reactions on their moving component due to the flow. These parasitic flow forces are undesirable and can be minimized by good design. As they may cause or account for unexpected system performance they should be appreciated.

The origins and nature of flow forces can be appreciated by considering the simple spool configuration of Fig. 10.3. In the centred state, the supply pressure acts on the two internal spool land faces equally and the pressure forces on the spool sum to zero. If the spool is moved to the right, as shown, the left control port remains blocked while the right control port experiences flow. The left internal spool land face continues to experience stagnant fluid at uniform pressure. However, the fluid flowing at relatively high velocity through the right-hand port gains kinetic energy, and this can only be achieved at the expense of pressure (potential energy). Thus, in the region of the jet formed by the partly-opened orifice, the pressure acting on the internal spool land face is reduced as shown. The axial force due to pressure on the right-hand spool land is reduced below that on the left-hand and a net axial force exists on the spool. The axial flow force acts to close the valve. If not resisted, it will do so. With manually controlled valves, the manual operator, possibly aided by a detente mechanism, usually can overcome the flow force and hold the valve open. It may become a problem with large valves. Flow forces are more significant with electrohydraulic valves, as their axial operating force is limited. The problem is worst with electrohydraulic servovalves, because of their sensitivity and the accuracy of flow required of them. There are spool-port design techniques which, though they add to expense, reduce flow force magnitudes.

Most of the major texts on hydraulic control systems provide details (e.g. Refs. 10.7, 10.8) allowing calculation of steady-state flow forces. Dynamic flow forces are less readily assessed.

Flow forces in valves are rarely of significance in predictive simulations of hydraulic control systems. Where necessary allowances can be made.

10.5 HYDRAULIC LOCK

Piston pumps and motors and spool-type valves rely only on the close fit of piston to bore to minimize leakage past the piston due to axial pressure drop. The fine clearances needed demand near-perfect geometry of both piston and bore, which is a main reason why hydraulic piston and spool components are expensive to produce.

In operation, pistons and spools experience a number of forces, including

- pressure area force in their axial direction,
- inertia force due to their own mass,
- viscous drag force, in their axial direction, due to viscous leakage flow in the clearance between piston and bore, and due to viscous shear of the oil film due to axial motion of the piston,
- flow forces, as previously discussed,
- mechanical friction due to any influence which causes the piston and bore to be in contact instead of the piston "floating" in the clearance oil film.

The first four of these effects are inherent and are well understood. Single-ended piston devices (pumps, motors, some pressure control valves) experience large but expected end pressure forces which are mechanically reacted in the device. Spool pistons are usually symmetrical so that the net axial pressure force is zero. These devices have the potential to be operated by a miniscule axial force, which has only to overcome the inertia and the viscous drag forces. A piston 1 cm diameter, 2 cm long, with diametral clearance 0.01 mm, and with axial pressure drop of 20 MPa experiences a drag force of only about 2 N . Valve flow can induce axial and radial forces on valve spools, inhibiting their performance and response, as

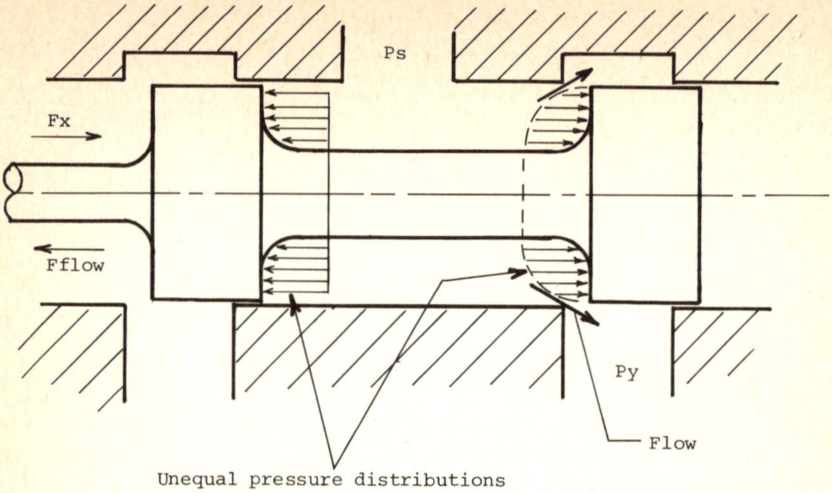

FIG. 10.3 ILLUSTRATING THE ORIGIN OF FLOW FORCE

a. Pump or Motor Configuration

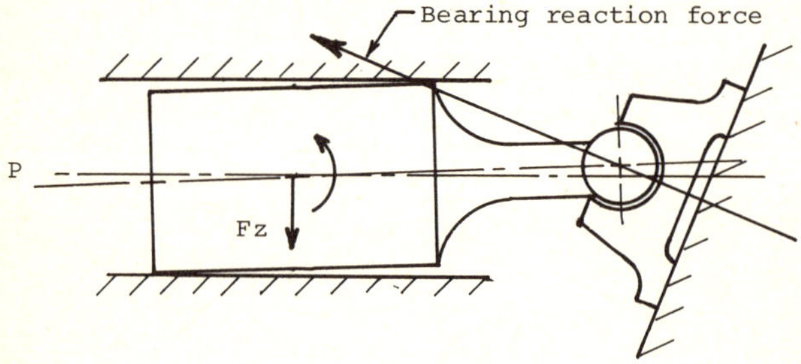

b. Tilting due to Unsquare or Non-Concentric Spring Support

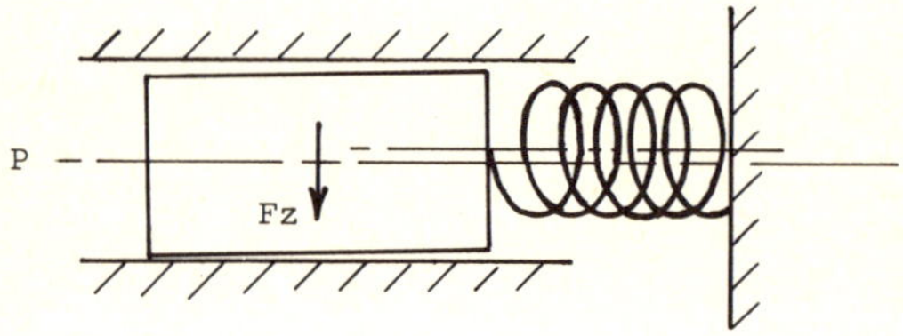

c. Bent Spool

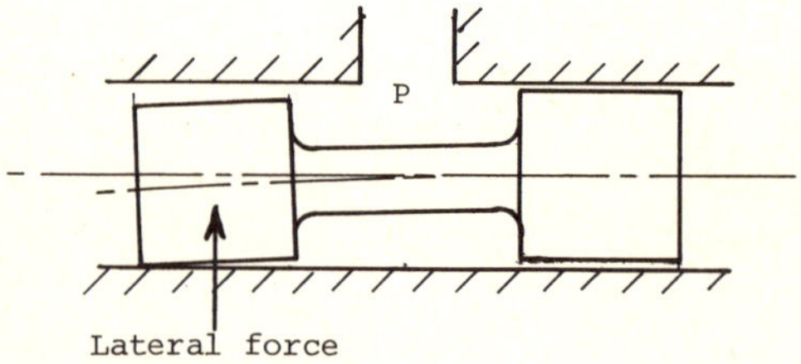

FIG. 10.4 SOME CAUSES OF UNSYMMETRICAL CLEARANCES IN WORKING PISTON COMPONENTS

separately discussed. Such flow forces can be over 100 N .

Mechanical friction requires that a lateral force exists to force the piston into contact with its bore. Then, any attempt to move the piston axially must overcome the resulting friction, and the force required to induce axial motion would include a component

$$Ffx = \mu \cdot Fz$$

where μ is coefficient of friction between piston and bore

Fz is the net radial force producing contact between piston and bore

x denotes axial direction

z denotes lateral direction.

With piston pumps and motors, the eccentric nature of the drive acts to induce piston-bore contact as the piston is forced against the bore (Fig. 10.4a). With pumps, the high speed reversing motion of the piston allows little time for lateral and tilting displacement of the piston to fully develop, and also encourages good hydrodynamic lubrication to be sustained. It takes time to squeeze out a film of mineral oil. In any case, the large power input to a pump completely overwhelms its piston friction effects, and piston friction is reduced to a negligible effect as far as dynamic response of the driven load is concerned. This is not so with hydraulic motors unless they too are run at reasonably high and constant speed. The well-known tendency of hydraulic piston motors to stick or to be jerky at stall or slow speeds is at least partly due to friction deriving from lateral piston displacement and tilt arising from the eccentricity of the drive.

Most spool type control valves are intended to be in hydrostatic balance, requiring minimal forces to move the spool in its bore. Yet, a spool which is free in its bore when no pressure exists in it can become hard to move under pressure. Quite apart from the flow forces, two other possible causes for this kind of valve sticking are contamination lock and hydraulic lock. Contamination lock will be discussed in Section 10.6. The remainder of the present section will introduce that parasitic phenomenon, hydraulic lock.

Fig. 10.5a shows a piston lying centrally in its bore with a high pressure P1 at one end and a lower pressure P2 at the other end. Ideally, one would imagine that the pressure distribution in the clearance is linear as shown, and that it is the same along all axial generators on the piston surface. That is, the axial pressure distribution at $\theta = 0$ is the same as that at $\theta = \pi$ and at all other values of θ , and the radial forces generated on the piston surface sum to zero, leaving the piston floating freely in its surrounding oil film.

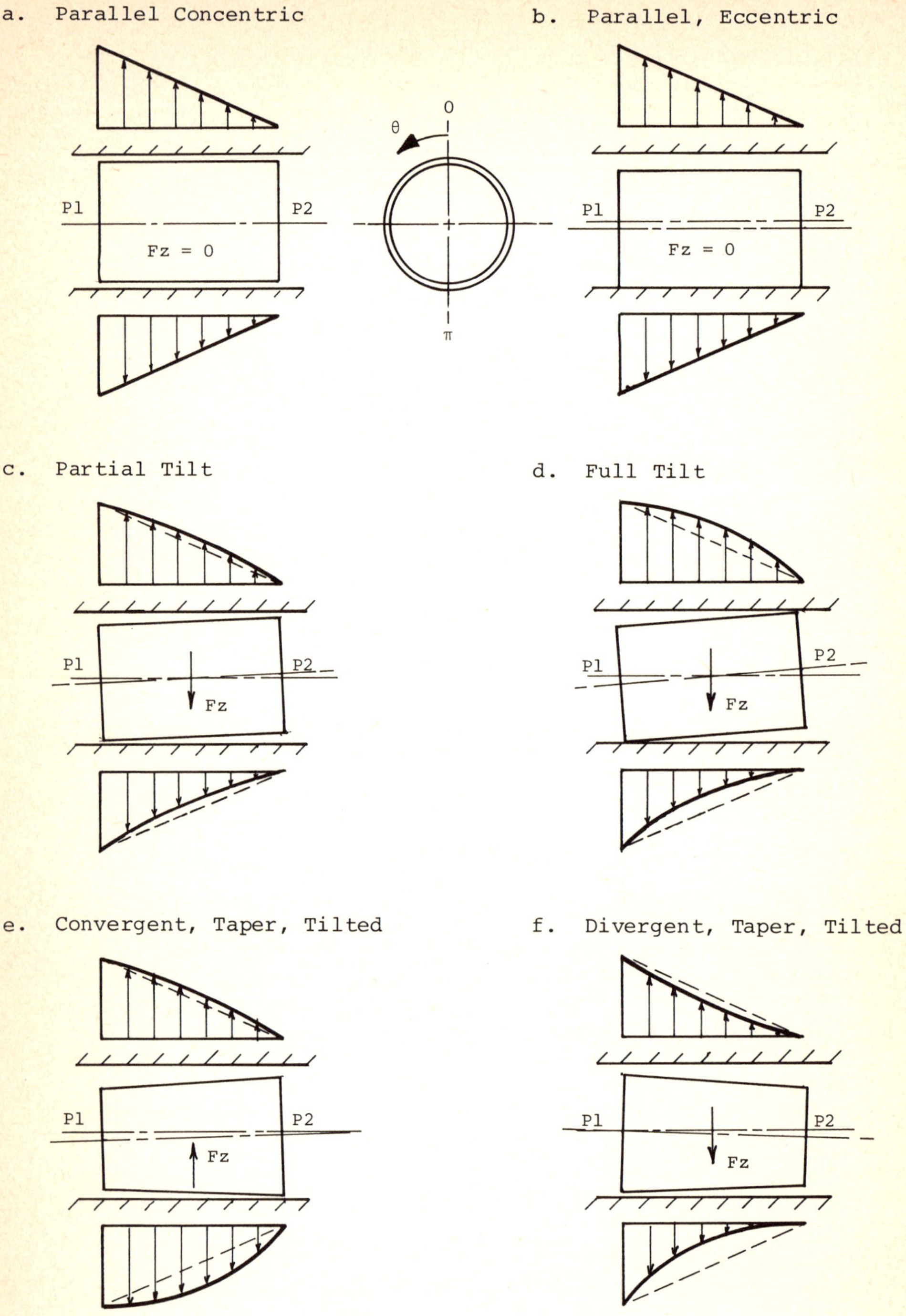

FIG. 10.5 CLEARANCE SHAPES AND PRESSURE DISTRIBUTIONS FOR VARIOUS PISTON-BORE CONFIGURATIONS

But what if the axial pressure distribution is not the same at different values of θ ? For example, Fig. 10.5c illustrates a humped pressure distribution at $\theta = 0$, and a sagging pressure distribution at $\theta = \pi$. Both still start, as they must, from P1 and end at P2 . The area under a pressure distribution represents a radial force generated on the piston. Thus, the $\theta = 0$ distribution exerts a downward force on the piston larger than the upward force of the $\theta = \pi$ distribution, and the piston is forced downwards towards contact with the bore. Frictional resistance against axial motion can then be expected.

Friction between piston and bore arising from uneven axial pressure distribution around a piston is known as *hydraulic lock*.

Under what conditions can unequal axial pressure distributions π radian apart arise? And under what conditions will the net lateral force of such pressure distributions cause decentring of the piston towards contact with the bore? These questions have been studied fairly extensively, for example in Refs. 10.9, 10.10, 10.11. If for any reason, the *actual* axial clearance between a piston and its bore is diverging in the direction of axial pressure drop, hydraulic lock can develop.

The attitude of a piston in its bore can rarely be guaranteed. Even in the ideal case of Fig. 10.5a, why should the piston be concentric with its bore? Its own mass can cause it to sink into the clearance oil film. The piston's supporting mechanism may cause it to tilt (Fig. 10.4a and b). A slight banana bend in a long spool can cause the situation of Fig. 10.4c. A geometrically perfect piston might have any of the attitudes shown in Fig. 10.5a to d. Also, geometric perfection of piston and bore (soundness, parallelism, straightness) is impossible to achieve. Imperfections contribute to uneven clearances and possible uneven axial pressure distributions. Lack of parallelism, in either or both components, can generate the clearance situations illustrated on Fig. 10.5e and f.

Fig. 10.5 includes the pressure distributions likely to be associated with various piston-bore clearance configurations. Each indicates the net lateral force Fz which will be generated by the pair of diametrically opposite pressure distributions shown. For the parallel concentric and parallel eccentric cases (a and b) Fz is zero. For the tilted piston cases (c and d) Fz acts to decentre the piston towards contact with the bore. With the convergently-tapered clearance situation (e) Fz is finite but acts to centre the piston in its bore, a highly desirable situation which holds whether the piston is tilted or parallel in its bore. With the divergently-tapered clearance (f) Fz acts to decentre the piston whether or not the piston is tilted. The decentring lateral force necessary for hydraulic lock to develop will be present if there is a diverging axial clearance due to piston tilt or component tapering.

Improved manufacturing techniques and quality control reduce the incidence of hydraulic lock. Grooving of piston lands reduces the magnitude of hydraulic lock by ensuring equal pressure around the piston at each groove. The expansion of the bore and compression of the piston due to pressure at their high pressure end acts to produce the desirable convergently-tapering clearance situation. Care should be taken that spools and bores are not bent axially, as this can result in the tilted piston configuration.

Hydraulic lock, if it develops, can be slight or substantial. If it develops in pumps, relief valves, or manually-controlled valves, it is usually overcome by brute force. In valves with limited operating forces, such as electrically controlled valves, it can cause complete jamming and malfunction. Such valves sometimes use electrical dither of the spool to keep the spool in motion. This prevents static friction and acts to inhibit the development of hydraulic lock, which is a time-dependent phenomenon at its worst if susceptible components are left stationary under pressure.

It would be most difficult to allow for possible hydraulic lock in dynamic simulation models, although a constant F_z force could be calculated and included in particular circumstances. It is far better to take steps to ensure that hydraulic lock is not a problem.

10.6 CONTAMINATED FLUID

Physical contamination is always present in hydraulic system fluids. Oil newly supplied contains millions of particles per litre. A system with appropriate filtration will clean up and maintain its fluid to a level at which the system components will operate effectively and with acceptable life. Much work has been done on studying the nature of contamination, the levels and distributions present in working systems, the measurement and monitoring of contamination levels, the effects on performance of contaminated fluids, and the tolerance levels of components to contamination levels (Refs. 10.12, 10.13, 10.14).

Physically contaminated fluid can erode and wear components, degrading the characteristics of control orifices and increasing internal leakage. It can also cause direct physical malfunction of sensitive parts. For example, a sensitive electrical control valve can silt up and jam if left stationary under pressure in the presence of excessively contaminated oil.

Specification of a contamination level acceptable for a particular system is a complex matter. It is insufficient to merely specify a gravimetric level, i.e. so many milligrams of contaminant per litre. The sizes, size distributions, and the materials of the particulate are also important. So too are the materials and clearances of the operating system components.

Fig. 10.6 illustrates crudely some mechanisms which could lead to performance aberrations in piston components. Hard particles of sizes of the order of the diametral clearance, and of half the diametral clearance, would be highly undesirable as they could enter the clearance and lodge in it, especially if pressure was reduced after entry. Silting at the entry to a fine clearance can affect pressure distribution in the clearance and so give rise to hydraulic lock in sensitive components.

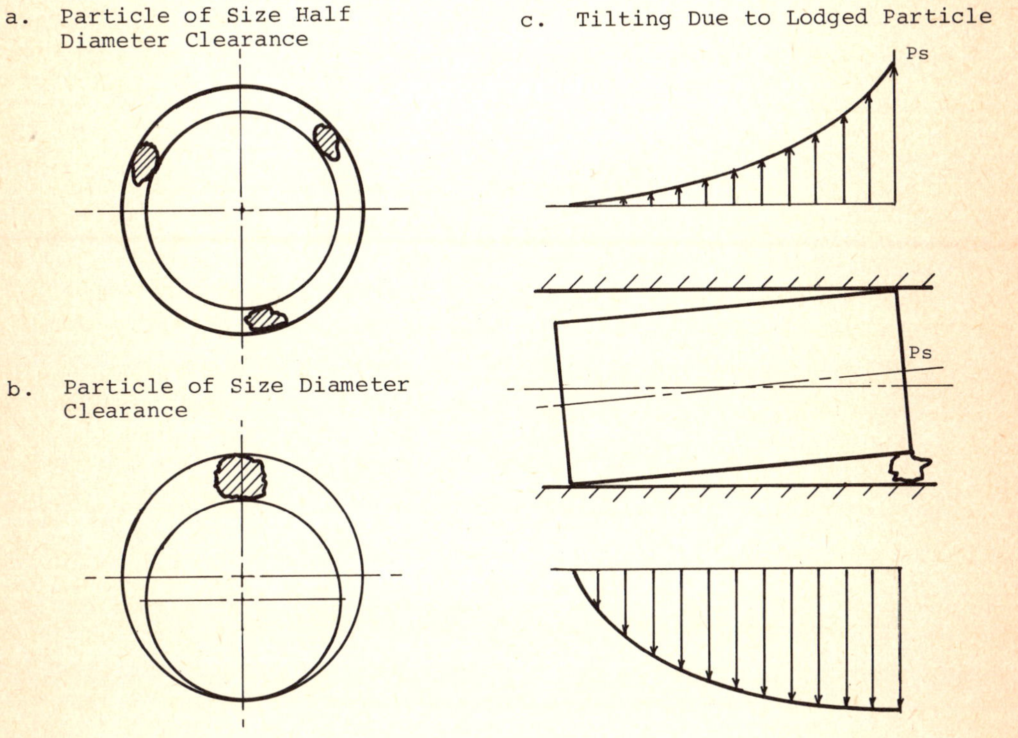

FIG. 10.6 SOME MECHANISMS FOR CONTAMINATION LOCK

The person concerned with the dynamic response and performance of hydraulic control systems should always be on the alert for aberrations in system performance which may be attributable to the effects of fluid-borne contamination. Such contamination will always be present. The idea is to keep it to levels at which the system performance will not be affected, and which will allow long and reliable life.

10.7 CONCLUSION

Some phenomena which might or might not manifest themselves in working hydraulic control systems have been discussed at an introductory level. There are other parasitic phenomena. Thermal lock is one. Thermal lock occurs if working components expand under the influence of temperature such that running clearances become impaired. If manifested, parasitic phenomena will affect system performance and response in a degradatory way. Most are not readily allowed for in dynamic system models. If a system's actual response does not correlate well with a predicted desired response, it could be due to one or more parasitic phenomena. For this reason, the designer-analyst of hydraulic control systems should have familiarity with them.

The phenomena discussed have been related mostly to piston configurations. They may apply also to non-piston components.

10.8 REFERENCES

10.1 McCloy, D., Some Causes and Effects of Cavitation in Oil Hydraulic Equipment, Proc. Conf. Fluid Power 1970, Fluid Power Soc., Melbourne, 1970.

10.2 Pearsall, I.S., Cavitation, Mills and Boon, London, 1972.

10.3 Wang, P.K.C. and Ma, J.T.S., Cavitation in Valve-Controlled Hydraulic Actuators, Trans. ASME, Jrl. App. Mech., V30, 1963.

10.4 McCloy, D., Cavity Formation in Valve Controlled Hydraulic Cylinders, Proc. I.Mech.E., V184, Part 1, N52, 1969-70.

10.5 Rogers, K.J., On the Origins and Effects of Hydraulic Backlash in Tandem Centre Valve Controlled Hydraulic Systems, Ph.D. thesis, Monash University, 1976.

10.6 Rogers, K.J. and Dransfield, P., A Study of Hydraulic Backlash by Simulation, Proc. Conf. on Control Eng., Inst. Eng. Aust., Melbourne, 1979.

10.7 Merritt, H., Hydraulic Control Systems, Wiley, 1967.

10.8 McCloy, D. and Martin, H., Control of Fluid Power, Longman, 1973.

10.9 Blackburn, J.F., Contributions to Hydraulic Control, Part 5, Lateral Forces on Hydraulic Pistons, Trans. ASME, V75, 1953.

10.10 Manhajm, J. and Sweeney, D.C., An Investigation of Hydraulic Lock, Proc. I.Mech.E., V169, N42, 1955.

10.11 Dransfield, P., Bruce, D.M., and Wadsworth, M., A General Approach to Hydraulic Lock, Proc. I.Mech.E., V182, Part 1, N27, 1967-68.

10.12 Brown, J.F.C., Contamination Control in Hydraulic Systems, Proc. 4th Intl. Fluid Power Symposium, BHRA, 1975.

10.13 Tessman, R.K. and Fitch, E.C., Contaminant Induced Wear Debris for Fluid Power Components, Proc. I.Mech.E. Technology Conf., Swansea, U.K., 1978.

10.14 Fitch, E.C., An Encyclopedia of Fluid Contamination Control for Hydraulic Systems, Fluid Power Research Center, Oklahoma State University, U.S.A., 1979.

CHAPTER 11

CONCLUSION

It is hoped that the Text has provided the reader with a reasonably deep appreciation of the potential and the use of bond graphs as a key link in improving his (read his or her) ability to predict the dynamic response of hydraulic control systems. I encourage the reader with the desire but not the confidence to effect such analyses, to try the bond graph approach. The reader already confident of his ability to perform such analyses on proposed systems on the drawing board, may well see bond graphs as just another modelling technique. I challenge such readers to dig in sufficiently to be able to properly try the bond graph approach to simulation. At the very least you will find it enlightening.

The Text has as its philosophy

- that conventional static design procedures allow proposal of a system which should carry out the desired functions,
- that it is insufficient merely that the required functions will be induced; the quality of the system actions and the time frame during which they occur are also important, often vitally so,
- that predictive dynamic response analyses should be carried out as part of the system design procedure,
- that bond graphs provide a key link in the confident development of dynamic models,
- that digital simulation is the way to extract response information from models,

- that procedures exist which allow reasonable optimization of system response,
- that the various phenomena *inherent* to the response of hydraulic control systems should be clearly appreciated,
- that there are phenomena *parasitic* to the response of hydraulic systems and that the system designer should be aware of these and of their possible effects.

It is not claimed that the Text is exhaustive. There are bond graph techniques, inherent and parasitic phenomena, optimization techniques, and related matters which have been barely referred to or neglected. There are however sufficient material and examples to allow the hydraulic system designer-analyst to investigate many systems, and References from which he can extend his knowledge when ready. The Author claims no omnipotence in the area of dynamic performance of hydraulic control systems —— simply a reasonable knowledge and experience of it which has been significantly enhanced by his adoption of bond graphs.

APPENDIX 1

STATIC DESIGN APPROACH

INTRODUCTION

A considerable number of books provide information and techniques for the design of hydraulic control systems. Some are included in the References of this Appendix. The designer rarely needs to design a major component. His task is to conceive a strategy appropriate to the particular application, to represent this strategy on a circuit diagram, and to select the components for his system from a wide range of commercial stock most of which will be available on-the-shelf. There may be delays in supply if the shelf happens to be in another country!

It has been assumed that the readers of this Text are reasonably familiar with hydraulic control systems and to some degree with their design. This Appendix will only outline a basic approach for arriving at a hydraulic system capable of performing a specific task.

POINTS TO BE CONSIDERED

- Appreciation of the task to be performed
  - type, size, mass, shape of load(s) to be driven;
  - motions required; type, size, speed, frequency of changes and reversals;
  - location of the task; factory, field, airborne, dockside, mobile, etc.;
  - working environment; dust or dirt levels, humidity, temperatures, noise levels, space limitations.

- Mode of operation and control
  - manual or automatic control;

. continuous or intermittent operation;
. how much full-power operation.

- Safety requirements
  - protection of operators and others;
  - protection of the system;
  - adjacent machinery or equipment.

- Reliability
  - life of system;
  - likely maintenance requirements;
  - access to systems;
  - ease of maintenance;
  - availability of maintenance expertise and facilities;
  - operator ability.

- Cost of the system
  - capital cost;
  - operating cost;
  - maintenance cost.

- Other factors
  - will operating noise be a problem?
  - does the client have any personal whims or requirements? Layout, preferred maker of components, compatibility with existing system, etc.;
  - Size, space, mass, or layout limitations.

AN APPROACH

1. Establish the functions, motions, and cycles required of the system.

2. Propose a hydraulic circuit diagram for the system showing only major components (actuators, control valves, pumps). This diagram constitutes a design strategy.

3. Appreciate any basic constraints placed on the system, e.g. maximum system pressure, pump drive speed.

4. Specify the maximum strokes required of linear actuators and/or the maximum speed required of hydraulic motors.

5. Assess the maximum force or torque required of each actuator, allowing for inertias, friction, and external forces or torques the drive will be expected to overcome.

6. Determine approximate actuator size(s) on the bases:

   . actuator piston area required = maximum force required/maximum pressure of system

   . volumetric displacement per radian of motor = maximum torque required/maximum pressure of system

7. Determine the maximum input flow requirement of each actuator on one of following bases:

   . inflow rate required = area of piston · maximum piston velocity required

   . inflow rate required = volumetric displacement per radian · maximum angular velocity required

8. Determine the approximate requirements of the main control valve(s), on the bases:

   . maximum flowrate required at the actuator(s) serviced by each valve;

   . maximum system pressure to be allowed;

   . type of control required, e.g. manual, electrical, closed-centre, etc.

9. Determine the approximate pump discharge rate required on the bases of:

   . the number of actuators to be driven simultaneously;

   . the maximum flowrate required at the actuators;

   . the maximum flow combination;

   . allowance for internal leakages of actuator(s), motor(s), and control valve(s).

10. Select commercial actuator(s) and control valve(s). Refine approximate pump discharge rate requirement accordingly.

11. Select a commercial pump, considering

    . speed at which pump is to be driven;

    . maximum pump discharge rate required;

    . maximum pressure to be allowed in the system;

    . volumetric efficiency of pump;

    . type of system and type of control (closed, open, tandem-centre, etc.);

    . choice between fixed displacement, variable displacement, over-centre variable displacement, pressure compensation, automatic power limiting and control features, etc.;

    . possible use of an accumulator.

12. Decide on oil filtering strategy. Select filters.

13. Decide if any intermediate valving or control circuits are required and select suitable components.

14. Decide hydraulic line or hose sizes (length, internal diameter, strength).

15. Update the circuit diagrams.

16. Estimate the maximum temperature rise in the system oil due to working of the system.

17. Decide on reservoir size and if oil cooling is required.

18. Select the primary drive for the pump.

19. Allowing for pressure drops, internal leakages, and for the actual sizes and ratings of the commercial items, refine all calculations.

20. Check that the system meets the action and performance requirements.

21. Estimate the cost of system components and of building the system.

22. Optimize the system and component selection on the bases of:

    . minimum cost;
    . adequacy of performance;
    . minimum number of components.

23. Finalize the design, component selection, and the estimate of cost.

The procedure outlined requires use of static or equilibrium relationships and component manufacturers' data. There can be no certainty as to the actual dynamic response and performance of the proposed system. Dynamic analysis of the type which is the subject of this Text should begin after Step 11.

REFERENCES ON BASIC DESIGN

Guillon, M., Hydraulic Servo Systems: Analysis and Design, Butterworths, 1969.

Yeaple, F.D. (Ed.), Hydraulic and Pneumatic Power and Control, McGraw-Hill, 1966.

Merritt, H., Hydraulic Control Systems, Wiley, 1967.

Kibble, J.D., Basic Factors in the Design of Hydraulic Circuits, Proc. I.Mech.E., V.184, Part 1, N.1, 1969-70.

Prokes, J., Hydraulic Mechanisms in Automation, Elsevier, 1977.

Khaimovich, E.M., Hydraulic Control of Machine Tools, Pergamon, 1965.

APPENDIX 2

SI CONVERSION FACTORS

The USA is the major industrial nation not yet fully using the SI system. The following table allows conversion of units commonly used in that country's fluid power industries to be converted to SI.

Quantity	Basic SI Unit	Common USA Unit	Basic Conversion	Alternatives
Basic Units				
Length	metre (m)	inch (in)	in/39.37 = m	in · 25.4 = mm in · 2.54 = cm ft/3.28 = m
Mass	kilogram (kg)	pnd.mass (lbm)	lbm · 0.4536 = kg	slug · 14.6 = kg
Time	second (s)	second	1	
Current	ampere (A)	ampere	1	
Temp. (gauge)	deg.cel. (C)	deg.fah. (°F)	(°F − 32)/1.8 = C	
Temp. (absol.)	deg.kel. (K)	deg.Rank. (°R)	°R/1.8 = K	
Derived Units				
Area	m^2	in^2	$in^2/1550 = m^2$	$in^2 \cdot 6.452 = cm^2$
Volume	m^3	US gallon	US gal/264 = m^3	US gal · 3.785 = ℓ
Density	kg/m^3	lbm/ft^3	$lbm/ft^3 \cdot 16.02 = kg/m^3$	
Force (also weight)	newton (N)	lbf	lbf · 4.448 = N	
Torque	N · m	lbf · in	(lbf·in)/8.85 = N · m	
Rot.inertia	$kg \cdot m^2$	$lbm \cdot in^2$	$(lbm \cdot in^2)/3413 = kg \cdot m^2$	
Pressure	pascal (Pa) Pa = N/m^2	lbf/in^2 (psi)	psi · 6895 = Pa	psi/145 = MPa psi/14.5 = bar
Energy (also work)	joule (N · m)	ft · lbf	(ft·lbf) · 1.356 = J	
Power	watt (W) (J/s)	h/power (hp)	hp · 746 = W	hp/1.341 = kW
Elec.potent.	volts (E)	volts	1	
Elec.resist.	ohm	ohm	1	
Linear vel.	m/s	ft/s	(ft/s)/3.28 = m/s	
Linear acc.	m/s^2	ft/s^2	$(ft/s^2)/3.28 = m/s^2$	
Flowrate	m^3/s	USgal/min	(USg/min)/15840 = m^3/s	USg/min·3.785 = ℓ/min
Angle	rad	rad	1	rev = 2π rad
Ang.velocity	rad/s	rad/s	1	rpm · 2π/60 = rad/s
Frequency	hertz (Hz)	Hz	1	cycle/sec = Hz
Dynamic viscosity	$N \cdot s/m^2$, or Pa · s	centipoise (cP)	cP/E3 = Pa · s	
Kinematic viscosity	m^2/s	centistoke (cSt)	cSt/E6 = m^2/s	

Note: 10^{n} is shown as En
10^{-n} is shown as E-n

APPENDIX 3

DYNAMIC RESPONSE—A BIBLIOGRAPHY

There is a remarkably large number of English language papers concerned with the dynamic response and performance of hydraulic control systems. They appear in a wide variety of journals, with fairly regular contributions in the Proceedings of the Institution of Mechanical Engineers, the Proceedings of BHRA's International Fluid Power Symposia, the Bulletin of the Japan Society for Mechanical Engineers, and in a number of the ASME journals. The present Bibliography is quite limited, choices for it being based on whether a paper is directly referenced in the Text, whether it is available in accessible literature, whether it is clearly relevant to the work of the Text, and an attempt to provide the names of persons working in the area. Some excellent older references are not included as they are referred to in later papers.

Books

Blackburn, J.F., Reethof, G., and Shearer, J.L., Fluid Power Control, M.I.T. Press, 1960.

Merritt, H.E., Hydraulic Control Systems, Wiley, 1967.

Guillon, M., Hydraulic Servo Systems: Analysis and Design, Butterworths, 1969 (translation from French).

McCloy, D. and Martin, H.R., The Control of Fluid Power, Longman, 1973.

Prokeš, J., Hydraulic Mechanics in Automation, Elsevier, 1977.

Viersma, T.J., Analysis Synthesis and Design of Hydraulic Servosystems and Pipelines, Elsevier, 1980.

Papers, Theses, Articles

Rausch, R.G., The Analysis of Valve Controlled Hydraulic Servomechanisms, Bell Syst. Tech. Jrl., V38, 1959.

Wang, P.K.C., Mathematical Models for the Time Domain Design of Electrohydraulic Servomechanisms, A.I.E.E. Trans. Applic. and Industry, V80, 1961.

Wang, P.K.C. and Ma, J.T.S., Cavitation in Valve Controlled Hydraulic Actuators, Trans. A.S.M.E. Jrl. App. Mech., V30, 1963.

McCloy, D. and Martin, H.R., Some Effects of Cavitation and Flow Forces in the Electrohydraulic Servomechanism, Proc. I.Mech.E., V178, Pt.1, 1963-64.

Davies, R.M. and Lambert, T.H., Transient Response of a Hydraulic Servomechanism Flexibly Connected to an Inertia Load, J. Mech. Eng. Sci., V6, 1964.

Dransfield, P. and Bruce, D.M., Leakage Flow Rate Past Pistons of Aℓ Hydraulic System Components, J. of Aircraft, A.I.A.A., V5, N2, 1968.

Healey, A.I. and Stringer, J.D., Dynamic Characteristics of an Oil Hydraulic Constant Speed Drive, Proc. I.Mech.E., V183, 1968-69.

Montgomery, J. and Lichtaravicz, A., Asymmetrical Lap in Hydraulic Servomechanisms, Proc. I.Mech.E., V183, Pt.1, 1968-69.

Lepp, R.M., A Study of the Time Optimized Hydraulic Servomechanism Operating Under Cavitation Conditions, Ph.D. Thesis, U. of Saskatchewan, Canada, 1968.

Takenaka, T. and Urata, E., The Dynamic Characteristics of Oil Hydraulic Control Valves, Bull. J.S.M.E., V12, 1969.

Burrows, C.R. and Webb, C.R., The Design of Fluid Power Systems Using Analogue Computer, Trans. Inst. Meas. & Control, V2, 1969.

Haines, D.F. and Davies, R.M., Harmonic and Transient Behaviour of an Electro-hydraulic Servomechanism Test Rig, Control, V13, 1969.

Nikiforuk, P.N. and Tsai, S., Detailed Analysis of a Two-Stage Four-Way Electro-hydraulic Control Valve, J. Mech. Eng. Sci., V11, N2, 1969.

Urata, E., The Ramp Response of a Loaded Hydraulic Servomechanism, Bull. J.S.M.E., V13, 1970.

Green, W.L. and Crossley, T.R., An Analysis of the Control Mechanism Used in Variable Delivery Hydraulic Pumps, Proc. I.Mech.E., V185, 1970-71.

Schlösser, W.M.J., A Contribution to the Analysis of the Characteristics of Hydraulic Valves, Hyd. and Pneu. Power, V24, N7, 1971.

Ulrich, H.J., Some Factors Influencing the Natural Frequency of Linear Hydraulic Actuators, Intl. of Mach. Tool Design and Research, V11, 1971.

Armstrong, P.J. and McCloy, D., Optimization of a Hydraulic Servo Using On-Off Controllers, Proc. 2nd Intl. Fluid Power Symp., BHRA, 1971.

Alpay, S.A., The Influence of Dynamic Effects in Discharge Coefficients in Spool Valves, Proc. 2nd Intl. Fluid Power Symp., BHRA, 1971.

Knight, G.C., McCallion, H., and Dudley, B.R., Connection Capacitance Effects in Hydrostatic Transmission Systems and Their Prediction by Mathematical Model, Proc. I.Mech.E., V86, 1972.

Bowns, D.E. and Worton-Griffiths, J., The Dynamic Characteristics of a Hydrostatic Transmission System, Proc. I.Mech.E., V186, 1972.

Keating, T. and Martin, H.R., Mathematical Models for the Design of Hydraulic Actuators, Trans. Inst. Soc. of Am., 1972.

Smith, C.K., Sebesta, H.R., and Bose, J.E., Stabilization of a Hydromechanical Steering System, SAE Paper 720791, 1972.

Unruh, D., A Standard Format for Mathematical Models of Fluid Power Systems, Proc. 26th N.C.F.P., 1972.

Armstrong, P.J. and McCloy, D., Design of High Speed Hydraulic Position Servomechanisms Using Optimization Techniques, Proc. I.Mech.E., V187, 1973.

Young, M.Y., Digital Simulation of Hydraulic Control Systems, M.Eng.Sc. Thesis, Monash University, Australia, 1973.

Barnard, B.W., The Dynamic Response of a Hydraulic System, M.Eng.Sc. Thesis, Monash University, Australia, 1973.

Chiappulini, R., A Review of Hydraulic Drives and Servodrives —— Some Actual Problems, Proc. 3rd Intl. Fluid Power Symp., BHRA, 1973.

Martin, K.F., Design Approximations for Hydraulic Servos with Non-Symmetrical Oil Volumes, Meast. & Control, V6, 1973.

Muto, T. and Hattori, T., A Study on the Unstable Phenomena in Hydraulic Driving System and Direct Servosystem, Bull. J.S.M.E., V17, 1974.

Schlösser, W.M.J., A Contribution to the Study of Analogies of Power Transmissions in Machines, Proc. I.Mech.E., V188, 1974.

Helgestad, B.O., Foster, K., and Bannister, F.K., Pressure Transients in an Axial Piston Hydraulic Pump, Proc. I.Mech.E., V188, 1974.

Dreymuller, J., Pilot-Operated and Directly Actuated Pressure Control with Variable Delivery Axial Piston Pumps, Proc. 4th Intl. Fluid Power Symp., BHRA, 1975.

La Brooy, R. and Dransfield, P., On the Time Response Optimization of Hydraulic Control Systems, Proc. 4th Intl. Fluid Power Symp., BHRA, 1975.

La Brooy, R., On the Optimization of Hydraulic Control System Response, M.Eng.Sc. Thesis, Monash University, Australia, 1975.

Dransfield, P. and La Brooy, R., Hydraulic Systems with Precision Reflexes, Machine Design, 20 May 1976.

Parker, G.A. and Desjardins, Y.C., Identification of an Electrohydraulic Servo Operating in the Underlap Region of the Valve, Proc. Conf. Fluids in Control and Automation, BHRA, 1976.

Martin, D.J. and Burrows, C.R., The Dynamic Characteristics of an Electrohydraulic Servovalve, Trans. A.S.M.E., J.D.S.M.C., Dec. 1976.

Chong, F.K., Effects of Different Pressure Relief Valves on Dynamic Response of Hydraulic Control Systems, M.Eng.Sc. Thesis, Monash University, Australia, 1977.

Barnard B. and Dransfield, P., Predicting Response of a Proposed Hydraulic Control System, Using Bond Graphs, Trans A.S.M.E., J.D.S.M.C., V99, Series G, 1977.

Le Vert, F.R., Dynamic Analysis of a High-Speed Electrohydraulic Transient Rod Drive System, A.S.M.E. Paper 77-WA/Fi CS-8, 1977.

Martin, K.F., Ong, C.S., Tee, B.L., and Little, E., Dynamic Analysis of an Electrically-Operated Pressure Control Valve, Proc. 5th I.F.P.S., BHRA, 1978.

Mirakawa, S., The Effect of Flow Force on Frequency Characteristics of Hydraulic Motor Driven by Spool Valve, Bull. J.S.M.E., V21, N154, 1978.

Svoboda, J., Analogue and Digital Modelling in the Design of a Hydraulic Vehicular Transmission, Proc. I.Mech.E., V193, 1979.

Dransfield, P. and Teo, M.K., Using Bond Graphs in Simulating an Electrohydraulic System, J.Franklin Inst., V303, N3, Pergamon, 1979.

Yamaguchi, A. and Ishikawa, T., Characteristics of Displacement Control Mechanisms in Axial Piston Pumps, Bull. J.S.M.E., V22, 1979.

Akashi, H., Nakagawa, T., and Osumi, T., Analysis of Vibration in Hydraulic Drive System, Bull. J.S.M.E., V22, 1979.

Davies, D. and Stecki, J.S., Hydraulic System Analysis —— Modelling of Fluid Transmission Lines, BFPR Jrl. V14, N3, Oklahoma State University, 1980.

Stecki, J.S. and Cuthbertson, S.C., The Effects of Changeover to Fire Resistant Fluids on Hydraulic Performance, Fire Resistant Hydraulics Jrl., V1, N2, pp 167-173, FPRC, Oklahoma State University, 1981.

Ramachandran, S. and Dransfield, P., Interaction in a Two-Channel Loaded Electro-hydraulic Drive, Proc. 6th Intl. Fluid Power Symp., BHRA, 1981.

Backe, W., DSH Program System for Digital Simulation of Hydraulic Systems, Proc. 6th Intl. Fluid Power Symp., BHRA, 1981.

Bowns, D.E., Tomlinson, S.P., and Bonson, L.A., Progress Towards a General Purpose Hydraulic System Simulation Language, Proc. 6th Intl. Fluid Power Symp., BHRA, 1981.

INDEX